Springer Biographies

The books published in the Springer Biographies tell of the life and work of scholars, innovators, and pioneers in all fields of learning and throughout the ages. Prominent scientists and philosophers will feature, but so too will lesser known personalities whose significant contributions deserve greater recognition and whose remarkable life stories will stir and motivate readers. Authored by historians and other academic writers, the volumes describe and analyse the main achievements of their subjects in manner accessible to nonspecialists, interweaving these with salient aspects of the protagonists' personal lives. Autobiographies and memoirs also fall into the scope of the series.

More information about this series at http://www.springer.com/series/13617

Douglas Jenkinson

Outbreak in the Village

A Family Doctor's Lifetime Study
of Whooping Cough

 Springer

Douglas Jenkinson
Nottingham, UK

ISSN 2365-0613   ISSN 2365-0621   (electronic)
Springer Biographies
ISBN 978-3-030-45487-6   ISBN 978-3-030-45485-2   (eBook)
https://doi.org/10.1007/978-3-030-45485-2

This Springer imprint is published by the registered company Springer Nature Switzerland AG
The registered company address is: Gewerbestrasse 11, 6330 Cham, Switzerland

It would seem natural to dedicate this book to the 744 men, women, children and their parents in the Keyworth practice between 1974 and 2013 who have been the core study group, and I do that with much gratitude, but there are others who were equally important. The ones, for instance, that I suspected of having whooping cough but didn't in the end include because they did not meet the criteria I set or because they had some other cause for their cough. They put up with my questioning, telephone calls, swabbing and blood testing willingly and even enthusiastically. Many have now moved on to live elsewhere, especially the children of course. All in all, these people, once my patients, must number thousands.

It was my pleasure to not only work in Keyworth as a family doctor but to raise my family there too. There could not have been a better environment in which to do both these things. It was, and still is, a typical English village community. Patients, friends and family intertwined in a way that blended seamlessly. My family and I thoroughly enjoyed it, and because we were part of the community, my few extra investigations were accepted. So this dedication includes the whole community of Keyworth and surrounding villages, and the friends we made there, that indirectly supported my endeavours.

Previous patient, Keyworth resident and friend Ian Bamford previewed the book and was inspired to draw a picture that made me laugh. It cheerfully disrespects me and the good people of Keyworth and a rather serious disease, but provides a welcome shaft of levity.

I also dedicate this book to family doctors who undertake or contemplate the difficult task of clinical research in their community. May they find the wherewithal. Family medicine and citizens need them.

Foreword

It is an honour to write this foreword for Dr. Douglas Jenkinson's book. He and I have been involved in the study of pertussis for about the same period of time although our backgrounds are completely different. I am a paediatric infectious diseases specialist and he is a general practitioner (GP).

My fellowship training was in virology; I got into pertussis by mistake. In 1976 there was an FDA contract to study reactions to whole cell pertussis vaccines (DTwP). A young faculty member in our UCLA Pediatric Department wanted to apply for the FDA contract and he asked me to be part of the pertussis vaccine research group. He asked me because I had a successful track record with other contracts with both FDA and NIH. We were awarded the contract and I have been involved in pertussis research ever since.

In 1982 I was awarded a scholarship that enabled me to go to the London School of Hygiene and Tropical Medicine (LSHTM) in the Epidemiology Program. In 1983 I received the MSc degree from the LSHTM. My thesis in the program was titled: 'The Epidemiology of Pertussis and Pertussis Immunization in the United Kingdom and the United States: A Comparative Study'. Of the 165 references in my monograph, two were notations by Douglas Jenkinson.

During the last 41 years (1978–2019) Dr. Jenkinson has 15 notations in the literature relating to pertussis. Of these notations 8 were full papers relating to pertussis. These were all keen observations relating to pertussis in his practice.

At the present time the pertussis research community has lost its way. This is because *Bordetella pertussis* causes infections in humans only. However, the majority of Bordetella researchers today do studies in animals (mainly mice) and the findings for the most part are not relevant to humans. Good epidemiological research is done at CDC and Public Health in the UK. However, we should all look at Douglas Jenkinson's studies in Keyworth, Nottinghamshire. He has documented four outbreaks, he has demonstrated DTwP vaccine efficacy, he has indicated false DTwP contraindications, he has identified 27 facts related to pertussis and he has documented pertussis in adults. He has produced a book that is fun to read. The book contents should be useful to practicing physicians, research scientists and the general public as well.

Distinguished Research Professor of Pediatrics   James D. Cherry
David Geffen School of Medicine at UCLA,
Los Angeles, CA, USA

Preface

Whooping cough (pertussis) is a disease of many facets, which every single person is threatened with from the moment of birth. It kills some babies and makes others very ill. Every year approximately one adult in 70 suffers a coughing illness caused by the whooping cough bacterium. It is the probable cause of about one in eight prolonged non-feverish coughs in adolescents and adults.

It may be severe, and recognised as clinical whooping cough, or mild and unrecognised. Either way it can kill a baby it is inadvertently passed to. It is so serious that every country has an immunisation programme to protect its children from it. In the UK it became officially recommended in 1957, but the vaccines used have been imperfect, so many individuals still catch it. But if there were no immunisation against whooping cough, it would probably be the commonest cause of death in young children.

It is a complex disease with a mysterious nature that is only now being unravelled. Anyone who thought whooping cough was a disease of the past will find this story a surprise. It is still very much with us but shrouded by misconceptions that this book will help to correct.

Previous generations lived under the constant threat of death from infectious diseases such as smallpox, tuberculosis, diphtheria, whooping cough and gastroenteritis, many of them coming in cycles every few years and repeatedly alarming the population. Antibiotics and vaccines have minimised these frightening events for three generations now. We were, however, reminded about previous times in the 1980s when HIV came along, and in this century the SARS-coronavirus and MERS-coronavirus sprang up but were seemingly suppressed by adequate isolation control measures. Another coronavirus, Covid-19, possibly by virtue of apparent asymptomatic transmission, escaped those methods of control and has painfully reminded the whole world of the damage infectious disease can do. We now know we need to be vigilant and prepared for new diseases as well as keeping up the fight against the old ones which many people have forgotten can be as bad or worse.

I was a family doctor in the Nottinghamshire village of Keyworth for 37 years and made a special study of whooping cough. The story can be told from a general practitioner (GP) perspective because the behaviour of the disease in a village is much the same as in the whole United Kingdom and most similar countries too. I started work as a GP in 1974, and by 2013, 744 carefully documented cases had helped me shed light on this enigmatic disease.

A pertussis vaccine brain damage scare, subsequently proved false, started in the UK in 1974 and caused many parents to reject the vaccine. Consequently, the disease returned after 25 years of control. Many babies died, and hundreds of thousands of children needlessly developed whooping cough. Recovery from that disastrous situation took a very long time, and the measures taken to restore an adequate vaccine uptake have had repercussions that are causing problems today.

The anti-vaccine sentiment that arose in 1974 was the first widespread rejection of a universal immunisation since the demonstrations against compulsory smallpox vaccination in Victorian times, and is being echoed today by antivaxx factions, mainly focussed on measles, mumps and rubella (MMR) immunisation. These concerns that are being voiced currently are much the same as 150 years ago. There are the same instinctive reactions of parents concerned for their children's welfare, and the same variety of other forces trying to direct or misdirect opinion. These elements are all here in this whooping cough story. There are lessons that can be learnt from these past events, but the best way to apply them today is debated. This story exemplifies the grave consequences of vaccine avoidance.

Pertussis vaccine acceptance eventually recovered and whooping cough almost disappeared from the official statistics in the UK and other developed countries in the mid-1990s, but I continued to see many cases. I knew that the reason the statistics were inaccurate was a consequence of doctors simply failing to recognise it, so I created a website in 2000 to help frustrated patients diagnose themselves. Over the following two decades it received millions of visitors.

Shortly after the millennium new testing methods for pertussis became available which made a difficult diagnosis much easier. At the same time as these tests became commonly used, the number of reported cases began to rise. In the UK it started in 2008 and peaked in 2012 when the highest level for 22 years was recorded. Similar rises occurred in Australia and the USA.

Most experts have called this large increase in reported cases a 'resurgence' of whooping cough, as if it had suddenly returned after an absence. The data I have accumulated in Keyworth over the last 40 years suggest that most of this 'resurgence', which is largely in teenagers and adults is no such thing. They have been there all along but simply not diagnosed. There does, however, appear to have been a recent small but real increase in whooping cough in babies, which is possibly related to the use of an acellular vaccine introduced around the millennium in many countries.

This is not the whole whooping cough story. Other health workers in different countries, or disciplines, would tell different ones. This is mine, from a 'whooping cough village' viewpoint, but taking full account of the research that has taken place over the years.

Some referencing has been included to give signposting for verification and to help those who want to dig deeper, but I have kept it to a minimum.

Some readers may want to understand something about the biology of *Bordetella pertussis* before starting the story. Appendix 1 will do that.

Nottingham, UK Douglas Jenkinson

Acknowledgements

So many people have contributed to this whole enterprise over more than forty years that it is impossible to name them all and I hope they will forgive me for missing so many out. I could never have done such a complete study without the help and encouragement of my medical partners. The late Manson Russell, Rowan Stevenson and Erl Annesley who took me into their partnership in 1974, and those who subsequently joined: Andrew Watson, Clive Ledger, Andrew Wood, Jill Langridge, Corinna Small, Jim Hamilton, Neil Shroff and Alan Carr. Health visitors Gwen Burgess and the late Jackie Pepper worked with me on the project for many years; it was only possible with their help. Nurse practitioner Liz Lewis took over their essential 'ear to the ground' role in later years.

I needed help with the various stages of writing papers. Special thanks must go to Professor David Hull who started me off. He was there at so many points and did more to help in the background than I realise even now. Dr. Ian Johnston repeatedly helped invaluably over the years, and Professor Angus Nicoll, when we collaborated.

Much of the work was necessarily done at home in evenings in front of the television or while watching my kids swimming on a Sunday morning. So thank you Joan (Ros), now Ros Manning, for enabling me to get on with my preoccupation so frequently. Our children, Nicola, Anna and Richard, were oblivious to it all then, but have all now caught up by helping me improve this book in various ways.

For producing the book from early drafts to the final version I need to thank the many friends who read through them and suggested improvements. Very special thanks are owed to MG who repeatedly scrutinised the text, made valuable suggestions and improved my grammar and syntax immeasurably. I also thank Jack Gardiner for his diligent scrutiny which helped me smooth out some clumsy prose.

Professor James Cherry of UCLA has given me enormous support and has kindly written a generous foreword. I am most grateful to him.

The immense knowledge and expertise of Professor Elizabeth Miller of Public Health England enlightened me about several crucial aspects of pertussis issues, and I am extremely grateful for her help and advice. Nevertheless, opinions expressed are my own and I take full responsibility for them.

I also thank everyone at Springer Nature who has helped build this book and especially Alison Ball in New York who has patiently and coolly coordinated the process.

Contents

List of Figures

List of Tables

About the Author

Douglas Jenkinson, DM FRCGP, was brought up on the Wirral Peninsula in north-west England and attended Calday Grange Grammar School. He graduated from Liverpool Medical School in 1967. After three years in junior posts in medicine, surgery, paediatrics, and obstetrics and gynaecology, he went to Zambia with his family where he spent three years doing general medical duties, obstetrics and gynaecology and neonatal paediatrics. There he discovered his love of research. He returned to the UK in 1973 to join a General Medical Practice partnership in rural Nottinghamshire where, in 1977, he investigated a large outbreak of whooping cough and was the first in 30 years to confirm the benefit of pertussis immunisation. He became a part-time lecturer in General Practice at Nottingham Medical School in 1979, and in 1988 with Professor Idris Williams, set up the first M.Med.Sci. course. He also researched and published papers on asthma and sat on the medical advisory committee to the Asthma Society and Friends of the Asthma Research Council. He contributed to textbooks on asthma and child health. He was awarded a doctorate for his whooping cough research in 1996 and continued to research whooping cough after he retired in 2011. He has a popular website to help patients with whooping cough get a diagnosis.

Chapter 1
Introduction

This book is a more or less chronological account of my investigation of all the cases of whooping[1] cough in the Nottinghamshire general practice I worked in from 1974 to 2011 and continued to case-find whooping cough in until 2013. Since then I have not been directly involved but the current doctors continue to diagnose whooping cough diligently, maintaining the consistency that started in 1974.

There were 744 cases in those 40 years. The study was not planned in advance. It happened almost by accident, but once I started I became so fascinated by this enigmatic disease that I could not stop.

I recorded as much detail as I could about each case and followed each one up until recovered. With the help of my medical partners and practice staff I was able to case-find amongst the whole practice population of over 11,000 and be confident that I had identified most of the clinically diagnosable cases that occurred. I believe this is the first and only time such a comprehensive long-term study of whooping cough in a defined community has been done. It has resulted in the publication of several scientific papers and shed light on the nature of the disease and the way its behaviour might have changed over the last four decades.

Most people know whooping cough as one of those diseases included in the primary immunisation of infants in their first few months of life, along with tetanus, diphtheria and polio, and nowadays several others too. The original 'triple' (*DTP*) vaccine, as it was known, had been a feature of the first year of life since the 1950s.

Thanks to the effectiveness of the vaccine most people don't get whooping cough and so it is confined to the small print of life in general. It causes fits of severe coughing that often end in the whooping noise that gives it its name. The whoop is a brief crowing sound that happens at the end of a coughing bout when air is rapidly sucked back into the lungs through a tight larynx. Vomiting frequently follows, and it can go on for several weeks with perhaps 20 or more such attacks in 24 hours.

---

[1] Some people pronounce the 'w' and call it wooping cough. In older literature it was written as 'hooping cough', which is probably how it is meant to be pronounced.

© The Editor(s) (if applicable) and The Author(s), under exclusive license to
Springer Nature Switzerland AG 2020
D. Jenkinson, *Outbreak in the Village*, Springer Biographies,
https://doi.org/10.1007/978-3-030-45485-2_1

These characteristics, and others I found, made it possible to distinguish whooping cough from other coughing illnesses relatively easily once you know how.

Without it being so distinctive a cough this study would not have been possible, because at the start in the 1970s there was no practical way of diagnosing it in general practice other than by its sound, and what is still particularly difficult is describing the sound of it to another person. Whooping cough creates a cough with a recognisable 'tune' which is instantly imprinted on the brain once it has been properly 'played'. But to know the 'tune' of whooping cough you need to have heard it. If you have had it yourself, or have had a child with it, you will forever be able to diagnose it in others better than most doctors.

Prior to 1957 when whooping cough vaccine became standard for all UK children, every doctor in the land and most parents could diagnose whooping cough because it was a common and familiar disease, the sound of it as well known as an emergency vehicle siren is today. In the developed world now, very many doctors are unable to diagnose whooping cough in that way because they have never knowingly heard it. Nowadays diagnosis is easily confirmed by blood or oral fluid testing for antibodies, or DNA testing of nose or throat swabs, but you can only do that if you suspect it, and to suspect it you still need to hear it or have it properly described.

Before the introduction of mass immunisation, whooping cough was a common disease in children and often the biggest cause of death in young ones. It is still a major killer in communities where immunisation is not available. It was a scourge of infants because they were often too feeble to survive the weeks of persistent coughing. They died from exhaustion, lack of oxygen, malnutrition, pneumonia or damage to the central nervous system from circulatory failure. In the 1930s approximately 100,000 cases of whooping cough were notified every year in England and Wales. One child in every hundred with it died from it; about 1000 deaths a year. Worldwide, 90% of the estimated 200,000 deaths a year from whooping cough are in underdeveloped countries. In the UK, even with sophisticated care, the fatality rate in infants is about one in a hundred. Fortunately, the actual number is small, but a sudden rise in 2012, probably related to an increased number of cases in older children, prompted the introduction of a pregnancy whooping cough vaccine booster in the October of that year. There were 14 deaths by the end of 2012, which was the peak, but none in 2017, and just one in 2018, reflecting the effectiveness of the booster.

Surprisingly little is understood about this disease that is unique to humans, and recent research reveals it to be unusual, complex and challenging to prevent or cure, but progress is being made.

Countries such as the USA and Australia have experienced an upsurge in reporting of whooping cough cases in the last decade or so and we started to see the same upward trend in reported cases in the UK in 2008. The reasons for this have only recently become clearer, but because this story is a journey of over 40 years, the possible explanations come later.

The nomenclature of this disease can cause confusion before we even begin. *Whooping cough* is the common English name for the characteristic coughing disease that is caused by *Bordetella pertussis* or *Bordetella parapertussis* bacteria. *Pertussis* is the correct medical term for the disease.

The story follows events chronologically, so some topics crop up more than once. I have been repetitive rather than refer the reader backwards and forwards.

Chapter 2
From Africa to England

I can trace my interest in whooping cough back to 1965 when I was a clinical medical student at Liverpool University. Myself and a group of about ten other students were on a ward round with paediatric consultant Dick Smithells who later became professor of Paediatrics and Child Health at the University of Leeds. We came to the cot of a child who had been admitted with whooping cough. He explained in detail aspects of the disease that were puzzling. There was the difficulty of knowing the precise cause. Was it always the bacterium *Bordetella pertussis* or might some virus also be responsible? Why did some children get it and not others, even if they were not immunised? Why did more girls get it than boys, when for most infectious diseases boys were most often affected? Why did it sometimes appear to lead to the chronic lung disease bronchiectasis? Why was there often a big increase in lymphocytes in the blood, the white blood cells more often associated with viral infections than bacterial infection? Why was it so hard to find the causative bacteria that were only discovered in 1906 by Belgians Jules Bordet[1] and Octave Gangou?

These mysteries were filed away into the back of my mind (along with many others!) as student years passed into junior doctor years and were occasionally reawakened when I saw cases of whooping cough in the isolation ward of Clatterbridge Hospital, where I spent three years rotating through house jobs, and later in a large district general hospital in Zambia where I worked as a General Duties Medical Officer (Fig. 2.1).

I discovered how much I enjoyed research during my time in Zambia. I spent three years there from 1971 to 1973 and found the experience exhilarating. There were two hospitals in the town of Chingola where I was based, and they were run by the Anglo-American copper mining company. My first six months were spent at a fee-paying hospital which catered principally for the very large expatriate, mainly

---

[1] Jules Bordet was awarded the 1919 Nobel Prize for Medicine. The bacterial genus Bordetella was named after him.

Fig. 2.1 Nchanga North Hospital, Chingola, Zambia 1973

European, community. It was similar to a cottage hospital[2] in the UK and integrated with a general practice service. Those six months served as an apprenticeship and introduction to my new environment before moving on to the bigger free hospital with 250 beds which provided all the medical services for the community of about a quarter of a million in that part of the Zambian Copperbelt.

There were so many unknowns about the best way of treating even the common diseases in that part of the world because so many diseases and problems co-existed in the same patient. The most frequent reasons for admission of children were measles, gastroenteritis and a variety of deficiency diseases. The death rate was sadly very high, often contributed to by underlying malnutrition, sometimes brought on by unnecessary and inappropriate bottle feeding, but often simply poverty, and ignorance about basic nutrition.

I was a 'General Duties Medical Officer' (*GDMO*). The only other grade was 'Specialist', but there were not many of them, and they came and went on mainly short-term contracts. One exception was the Irish international rugby player Jackie Kyle who was the surgical specialist. Modest, highly skilled, versatile and supremely personable, he seemed to have been there for ever and was our anchor, mentor and father figure. There was, from time to time, an obstetrician and gynaecologist. Chris

---

[2] Cottage hospitals were common in the early days of the NHS but have now mainly disappeared. They were small community units with beds and other facilities managed by local GPs and visiting consultants. Their quality was variable, but they were very popular.

Ruoss was there when I first arrived and introduced me to the kind of problems I would have to deal with that I had never encountered before. John Siddorn was a general physician who joined halfway through my contract. He had worked in Zambia before, so his experience was invaluable. He later went on to the University of Lusaka Medical School where he investigated a sudden rise in usually rare cases of cryptococcal meningitis that turned out to be the start of the worldwide AIDS epidemic.

We eventually got a proper specialist anaesthetist too. Brian Duffy was from Adelaide and was very keen to teach. He never tired of training as many of us as possible to give anaesthetics. John and Brigid Salt were two of his most enthusiastic pupils. They joined us fairly late on and had driven up from Capetown, a distance of 2000 miles, in a clapped-out old bread van. John went on to become a distinguished anaesthetist himself. We also had the services of a visiting consultant radiologist. When there was no specialist one of the *GDMO*s filled the gaps. Franco Trafficante became the anaesthetist, Terry Geddes became the physician, Colin Jones the paediatrician, and I became the obstetrician and gynaecologist and neonatal paediatrician. If Jackie Kyle went away Bruno Fernandez took over. Overseeing the whole service was Dr. Roy Digby, the Chief Medical Officer, whose job was mainly administrative.

This hospital at Chingola stimulated my sense of inquiry in medicine and I was able to get on with investigating and innovating, provided no extra funds were required! I never expected to find research so exciting. Questions were constantly arising about the diseases that we were seeing, and how they should be managed in this completely different environment. There were few textbooks to help and our training in the developed world seemed barely applicable in Central Africa. All we *GDMO*s had trained in the 1960s and I believe we had a more mature attitude towards ex-colonies than some of our predecessors who could be patriarchal and patronising. We wanted to help these developing independent countries with our skills as a way of showing support and respect, as well as gaining experience, and to be honest, earning a little more than we could in the UK. It was about this time that Lusaka University started turning out its own medical graduates. The days of the likes of us were numbered.

I completed several research projects in Chingola in those three years. Some were published at the time, some I had to work on and publish later. Some as proper papers and some as letters, because there was not enough time to do a comprehensive write-up.

The first I undertook was a determination of the incidence of symptomless urinary infection in pregnancy in the local population. This condition was believed to predispose to more serious kidney infection, which could in turn jeopardize a pregnancy, and in the developed world the rate was about four per cent. It turned out to be the same rate in Zambia.

The biggest project, which was later published in the *Journal of Tropical Medicine and Hygiene*, [1] was a radical method for preventing the very commonly found iron deficiency anaemia in pregnancy, using iron injections. That method was overwhelmingly successful compared with the standard alternative of iron tablets that were handed out but rarely taken.

I also made a bizarre observation in the maternity unit that is still I think unanswered today. It was something strange I noticed about patients suffering from eclampsia.

Eclampsia is a rare but serious complication of pregnancy that causes high blood pressure, fits and multiple organ damage in the mother and the death of the baby unless urgent measures are taken and the baby delivered. I saw quite a lot of cases because the antenatal clinics where pre-eclampsia (the early stages) would have been spotted were frequently not attended.

The patients were almost invariably black African or sometimes of mixed race. Most had had some primary education but very few spoke English. It struck me one day that all the patients I could recall with eclampsia were able to speak English. This seemed very strange. I tested this by finding the education level of all subsequent eclampsia patients and compared each with nine case controls from the same maternity ward for comparison. There was a strong association between eclampsia and the ability to speak English, which I published as a letter in the *British Medical Journal* (*BMJ*). I think the observation is still without a convincing explanation.

While in Africa I decided on a career in general practice and returned to the UK with my wife and young family to hopefully find a nice place to settle down.

There were not many GP jobs available at that time, and after looking at vacancies in Liverpool and Bedfordshire I really liked the feel of a semi-rural practice just south of Nottingham with a cheery down to earth atmosphere. I was fortunate enough to be taken on to replace a doctor who was leaving. The existing partners, Manson Russell, Rowan Stevenson and Erl Annesley, made me very welcome. It was rather like joining a family. In those days when partners shared 24/7 responsibility for all the patients 365 days a year, it was vital to be able to work cooperatively and amicably together and always pull your weight. It was certainly that way in the Keyworth practice and I counted myself very fortunate.

There was also a new Health Centre to work from in the middle of what was a recently much expanded village, alongside health visitors, district nurses and midwives, following the newly developed concept of a 'Primary Care Team'. This was very different from the situation that had prevailed previously, which was largely GPs working single handedly, often from a room of their own home converted into a consultation suite.

The practice looked after the village of Keyworth and several surrounding villages on the Nottinghamshire and Leicestershire border. There were about 11,500 patients in all, and I was responsible, in addition to normal GP services, for the Child Health Clinic, then run under the umbrella of the Nottingham Health Authority. I was very pleased about this because I was by then quite well experienced in paediatrics and had obtained the Diploma in Child Health, but as far as the patients were concerned, I was just the GP who was also the 'baby doctor'.

Changes were being proposed to improve children's medical services in the community. In 1976 the Committee on Child Health Services published 'Fit for the Future' [2], a plan to improve the care of children at a primary care level by

designating a doctor within each practice a 'GP Paediatrician'. This ambitious document was authored by distinguished paediatrician Donald Court and was thereafter
known as the 'Court Report'. When I read the report I felt encouraged, as I felt I
could fit such a role, but first I needed to improve my skills by receiving additional
training. Luckily I discovered that the enthusiastic and forward thinking professor
of Child Health at Nottingham's newly established Medical School ran a course to
train doctors in community paediatrics, so I applied to join it. I would need to take
a day out of the practice once a week during term time for a year, and of course, find
the necessary fee.

The professor in question was David Hull, later Sir David Hull and founder
President of the Royal College of Paediatrics and Child Health. David Hull gave me
much advice, help, support and encouragement in many of my academic endeavours over subsequent years. Without his help I doubt the Keyworth whooping cough
study would have happened. He was a mentor of great skill and kindness.

Getting time off for this course was difficult. My partners expected me to pull my
weight because if I was absent they would have to put in extra time to make up for
it. This was quite difficult because we had to provide 24-hour cover for our patients,
who expected us to turn up for medical problems at any time of day or night.

When we were 'on call', which was every fourth night and weekend, it was usual
to have to get out of bed at least once a night and often several times, and always be
there for the next day's work regardless. This was in addition to the normal Monday
to Friday and Saturday morning. We never received nor gave each other sympathy.
This was what we had signed up for.

Many patients had no telephone, and even fewer has access to personal transport,
so a call out would happen on the basis of a possibly garbled message passed by a
neighbour to whoever was taking the doctor's telephone calls. The compensation
was that after morning visiting was over, we were often free in the afternoon until
the start of evening surgery. So we managed to reach a compromise about my
course. I would do my normal morning surgery, but I would start it early and finish
it early, my partners would do my home visits for me, and I would get back from the
course in the late afternoon in time to do evening surgery. Perfect!

It was a good thing that in those days, patients were long-suffering. By and large
they didn't like the new appointment systems with receptionists often prying into
the reason for attendance. They were used to just turning up and waiting their turn,
no questions asked. So if my session at the university ran over, and I was half an
hour late starting surgery, nobody would generally bat an eyelid. If there was any
complaint the receptionist would mutter something about a difficult delivery out at
Willoughby-on-the-Wolds or some other distant village. Very often it was absolutely true. We justified this kind of thing by telling ourselves it was the only way of
improving the service. And so it was. Things work differently nowadays
fortunately.

An important element of the Community Child Health course I had enrolled for
was the need to complete an original research project related to the subject. The last
day of the course would be the day everyone presented their work. Fate was to

intervene and give me my project 'on a plate'. On that plate was an outbreak of whooping cough.

*****

For about three years roughly half of parents had declined whooping cough vaccine because of publicity about severe side effects. Nobody knew what the consequences were going to be. Some experts said nothing would happen, it was a problem of the past; others were worried it would come back in epidemic form. The answer came to Keyworth in August 1977 when a big outbreak started. Little did I know that these new cases were going to take my medical interests and career in an unexpected direction.

*****

References

1. Jenkinson D. Single-dose intramuscular iron dextran in pregnancy for anaemia prevention in urban Zambia. J Trop Med Hyg. 1984 Apr;87(2):71–4.
2. Court D. Report of the committee on child health services. Fit for the future. 1976.

Chapter 3
The Children's Clinic at Keyworth

Fig. 3.1 Keyworth Health Centre on Bunny Lane. Built in 1970 and demolished in 2007, it was replaced by Keyworth Primary Care Centre (Fig. 15.8). In the background is the house that until 1970 was the 'old surgery' and home of senior partner Dr. Manson Russell

When I had arrived back in the UK from Zambia in early December 1973, the country was in turmoil as a result of the 'three-day week' imposed by the Heath government to deal with the industrial action affecting coal production. I was completely unaware of the whooping cough vaccine controversy that was just starting to unfold, and was to cause problems lasting two decades, long after the three-day week had been forgotten.

Unnoticed by most people, in October 1973, a doctor from Great Ormond Street Children's Hospital had reported to the British Paediatric Association the cases of a series of children he believed had been damaged by whooping cough vaccine. It might have remained an issue kept within, and dealt with by the medical profession, had not a member of parliament learnt of it and was setting about preparing to bring it up in the House of Common at much the same time as I started work at Keyworth. A few weeks later this MP was to ignite the controversy that three years later impacted me, Keyworth and the country, in a big way. As the issue gathered momentum far away in London, I was getting to grips with a new and challenging job (Fig. 3.1)[1].

*****

One of the joys of being a GP is getting to know your own patients. It works two ways. You get to know their problems, they get to know you, and hopefully come to trust you. Both parties benefit tremendously most of the time from this. Because I or one of my three partners was accessible 24 hours a day and seven days a week in those days, there was true continuity of care. This is one of the main benefits of the NHS GP system, although the popularity of part-time working and the need to restrict working hours for health and safety reasons have eroded it in recent years. In those days we coped with long hours and chronic tiredness more easily because we did not work under such great pressure of high patient expectations or to such high standards as are required today.

The down side to this at the time was that, as a new doctor to the practice, I was replacing a previous doctor who had built up a good reputation and trust with patients and they had lost all that. A new doctor is automatically treated with suspicion and some mistrust. Only time and hard work can remedy it. Because part of the job delegated to me was child health care, which involved me seeing all the new babies, I had a shortcut to meeting parents and the children themselves, and they got to know me. Being the father of a young family myself at the time helped me empathise with the day to day difficulties of child care. This familiarity became very important in the whooping cough outbreak to come.

The 'baby clinic', as it was known at Keyworth Health Centre, started at 2 pm every Monday and was for the under-fives. It was run under the auspices of the Nottingham Health Authority, which meant that it was for everyone, not just those registered with the practice as patients, although in effect it was just the same,

---

[1] Pictured is the occasion of a gas leak in May 1990 when we all had to evacuate to the car park. Patients were so insistent on having a consultation they recounted their problems to all and sundry while we doctors hastily scribbled a prescription.

because we were the only practice, and anyone outside the area would go to a different clinic anyway.

'Organised chaos' would be a suitable description on most occasions as up to a dozen toddlers fought over the toys and tried to carry them off to a private corner that did not exist. The purpose of the clinic was to make a doctor and health visitor available to mothers of children under five years of age, in order to advise on child-care and development. The weighing of babies and young children was a large part of this and took place in a corner of the clinic waiting area. I had a small clinic room, also off this area, where consultations could take place confidentially. Formal medical checks of the children would be done at the ages of six weeks, six months, eighteen months and three years, to assess physical health and whether the children were progressing normally and achieving their 'milestones'. These checks had to be done by a doctor.[2] I also ran the immunisation clinic that replaced the baby clinic every fourth week and could involve up to 50 injections in a space of two hours. It was considerably noisier than normal clinics!

I was most fortunate to find myself working with a wonderful health visitor, Gwen Burgess. Gwen had been born and raised in Keyworth where her mother had been the local midwife. All the mothers loved her for her gentle style and empathy. She was also clinically shrewd and like many such people had an instinct for spotting when things were going wrong before most others noticed. She had trained in the days when to be a fully fledged health visitor you had be a nurse and midwife too (Fig. 3.2).

Gwen had watched Keyworth grow in the 1960s as hundreds of new houses were built on what was previously farmland. The population expanded many times over, and with it came new primary schools, a library, new shops, a comprehensive school and our health centre.

Most of the new people were young, with young children, or about to start a family. Teachers, nurses, young professionals of all sorts, and skilled tradesmen, most starting a new phase of life in a brand new home in a delightful part of the country. This was all highly relevant because they read their newspapers and kept up to date with current affairs, so in the years 1974 onwards when the media were reporting the possible association of whooping cough vaccine with brain damage, these parents took immediate notice. Consequently, the acceptance rate of whooping cough vaccine fell more rapidly locally than nationally.

Gwen had also watched the transformation of the medical service too. From GPs working single-handedly with little or no support except from their wives to more comprehensive 'group practices'. A new GP charter in 1966 had encouraged the formation of these 'group practices' with several GPs, employed staff, appointments, proper premises and better support and co-operation from hospital laboratory and x-ray departments.[3] Structured preventive child health surveillance could

---

[2] This task was later taken over by health visitors.

[3] General practice had become the 'Cinderella' of the NHS after it was created in 1948 but was becoming increasingly recognised as the backbone without which the rest might fail.

Fig. 3.2 Health Visitor Gwen Burgess making sure the polio drops have gone down in 1989

succeed in these new conditions and much progress followed, but there were instances of impropriety that make us cringe today. Gwen recalls that one of the first doctors in the Keyworth baby clinic was an incorrigible chain smoker, and Gwen's principal activity in the clinic was to quickly brush the ash off the examined babies as it fell from the smouldering embers between her lips.

Gwen still lives in the village. When I bump into her we reminisce about such things. Her contribution to the whooping cough study was invaluable, not least because she was able to pick out the sound of a child with whooping cough in the waiting room thirty paces away and out of sight, then whisk it away quickly to an isolation room to protect the other patients from infection.

As enthusiastic disease preventers, we medical staff at Keyworth were keen to keep our immunisation rates as high as possible. We were fortunate that the population was fairly stable and static, which meant that because almost the whole population was registered with us, anyone who failed to get immunised could be followed up and given the necessary jabs or polio drops (how it was given in those days). There were always a few families, usually deprived in one way or another, who were insufficiently organised to attend the clinics. If this happened repeatedly, I would usually make a home visit and do it there to make sure all were protected.

Chapter 4
Whooping Cough Vaccine

Of all the common childhood infectious diseases, such as mumps, measles, rubella, polio and chickenpox, whooping cough is by far the worst, whether judged by the severity of the illness or the number of deaths. Before there was immunisation against it, a high proportion of children would get whooping cough with varying degrees of severity.

The usual age to catch it was between three and six years and in general the younger you were, the worse you were affected. The usual disease was characterised by violent and exhausting bouts of coughing, frequently with vomiting and momentarily stopping breathing. Babies came off worst and frequently died of malnutrition, exhaustion, pneumonia and lack of oxygen leading to fits and brain damage. There were 100,000 cases notified annually in the UK with 1000 deaths every year in the 1930s and 1940s. Further back there were almost certainly more. The pattern is similar today in parts of the undeveloped world, although the WHO has brought about remarkable reductions in cases in recent years through its immunisation programmes.

There is no really effective treatment for the disease even today, but seriously affected babies can nowadays be supported with intensive care procedures, such as external oxygenation of the blood, that improve the chances of survival, and giving antibiotics for complicating infections. Very recently exchange transfusion has been tried with encouraging results. The only solution for reducing and controlling the disease at present is by immunisation of the whole population. In the future, a better vaccine may be able to eradicate the disease because it only affects humans.

Immunisation is something we all know something about. It is the most beneficial of all medical interventions and by and large it has had a good history. The general perception is that it is a good and sensible thing, particularly in the years following the introduction of a new vaccine when everyone can see the benefits. Unfortunately these can be forgotten with the passage of time.

© The Editor(s) (if applicable) and The Author(s), under exclusive license to
Springer Nature Switzerland AG 2020
D. Jenkinson, *Outbreak in the Village*, Springer Biographies,
https://doi.org/10.1007/978-3-030-45485-2_4

The first mass immunisations in the UK started in the 1790s against smallpox using cowpox material. After decades of experience and refinement it was deemed so effective that it was made compulsory in 1853, although the compulsion was eventually removed, but not until 1898.

Diphtheria and tetanus vaccines were developed in the 1930s. These bacteria damage us with a single toxin which make it much easier to produce a vaccine against than other bacterial infections. Diphtheria vaccine was not officially recommended for infants until 1942 and an anti-diphtheria publicity campaign was started to promote it. Subsequently it became a normal part of good childcare that mothers would have their babies immunised in the first year of life with the available vaccines, smallpox and diphtheria. Diphtheria had been a common and well-known killer prior to 1942. The number of cases subsequently fell dramatically (Fig. 4.1).

A vaccine combining tetanus with the diphtheria subsequently became available and came into routine use. In 1956 inactivated polio vaccine that was also given by injection came along and was welcomed enthusiastically in the wake of several summers with high numbers of paralytic polio cases.

Whooping cough vaccines were developed slowly. Several different types were being tried out in the 1930s and 1940s with varying degrees of success. They all consisted of large numbers of killed *Bordetella pertussis* bacteria. A killed whole cell vaccine was the only way a usable vaccine could be produced at that time because nobody knew what the harmful elements were that caused the disease, let alone extract them for a purer vaccine as had been done with diphtheria and tetanus.

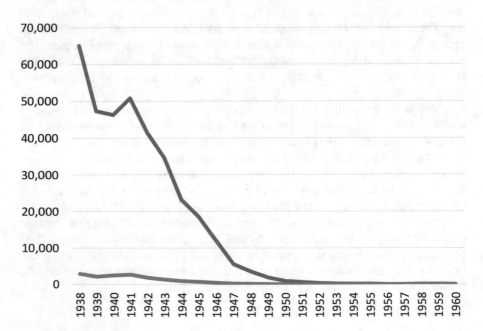

Fig. 4.1 Diphtheria in England and Wales showing the dramatic effect of immunisation on notifications (blue) and deaths (brown) 1938–60

By the 1950s there were whooping cough vaccines that provided about 80% protection to household contacts and had good safety profiles. That level of protection was considered worthwhile and had the potential to save many lives.

A 'triple' combination of diphtheria, tetanus and whooping cough vaccines was first licenced in the USA in 1948 and was available in the UK from 1952, but it was not officially recommended as national policy until 1957. Every local authority in the UK made its own decisions about vaccine deployment and they did it at different times in different places. Consequently, the statistics on uptake of these vaccines in this era are imprecise or absent and nobody knows exactly how many babies were given 'triple' vaccine between 1952 and 1957.

This uncertainty was not of great consequence for the monitoring of smallpox, tetanus, diphtheria and polio, whose incidences had plummeted dramatically and whose vaccine effectiveness was never in doubt, but it became a very important issue for whooping cough because the start of the drop in numbers of children with that disease either came before the introduction or after it, depending on whether 1952, when it became available, or 1957, when it was officially recommended, was taken as the critical date.

The national immunisation rate for triple vaccine from 1967 up to 1974 was steady at about 78%, although this figure may have been an underestimate because of the difficulty of producing accurate statistics at that time. There are no figures for the years prior to 1967 but the available evidence suggests a high uptake. We knew our immunisation rate in Keyworth was as high as 95% because the population was geographically discrete and stable, with permanent medical staff who were able to give a comprehensive service to the under-fives through the 'baby clinic'. In practice that meant that every baby received the 'triple vaccine' unless there had been a previous reaction to it, in which case the course would be completed with diphtheria and tetanus only.

There was a sudden change in 1974 when a big drop in whooping cough vaccine acceptance in Keyworth occurred as parents refused it following headline media reports about it causing brain damage. They opted for diphtheria and tetanus vaccine alone. It seemed a not unreasonable response, given that questions were being asked in parliament based on a report from a Great Ormond Street specialist. Several doctors I knew were also declining it for their own children.

The vaccine acceptance rate dropped rapidly all over the UK. In England and Wales for 1974 births it was 77%. For 1975 births it was 59%, and for 1976 births, 38%. Nobody knew what would happen next. It could be nothing if the experts who thought the vaccine ineffective were correct, or we could see a lot more whooping cough if the ones who believed the vaccine worked were right.

One of the problems for whooping cough is the way cases are counted. Unlike polio or diphtheria it is not a clear-cut disease and therefore falls victim to the weaknesses of the notification system. A law introduced in 1899 required doctors to notify to a proper authority, cases of disease that were on the notifiable diseases list, in order that potential outbreaks could be identified at an early stage, and control measures put in place. Smallpox, typhus and typhoid were amongst the first on the list. Whooping cough became notifiable in 1940, so statistics are only available from that date in the UK.

When the notification system was first introduced doctors could be fined for failing to notify a disease. Later this was changed to a payment for each case notified, sometimes quite well incentivising the process and at other times not, as the fee failed to match inflation. As can be imagined, the statistics thus obtained bore only a vague and variable relationship to the true incidence of the disease in question as many doctors regarded the process as bureaucratic and without medical relevance in many instances. Notifications of a serious disease such as tuberculosis would probably relate closely to the true incidence, whereas an ill-defined disease such as food poisoning inevitably underestimated the numbers.

These methods of counting made sense when the only way of making a diagnosis was by clinical judgement. With the enormous advances that have been made in the biomedical sciences, such methods have been superseded by more precise laboratory-based methods. In the case of whooping cough an enhanced surveillance system that started in 1994 requires health professionals to notify their local Health Protection Team (*HPT*) of a suspected case, by telephone or mail using the appropriate official form with detailed information about the patient. The *HPT* may then arrange for test samples to be taken from the patient, if possible and appropriate, and not already done. If the test is positive the patient is included in the confirmed case statistics gathered by the Public Health Laboratory for the district and passed to Public Health England (or other UK country) for inclusion in the national statistics. Thus, in the case of pertussis, dual statistics are kept, notified cases and laboratory confirmed cases.

The advent of a reliable blood testing method in 2002, and oral fluid in 2013, rather than the previous unreliable culture method, was a big step forward in improving confirmation and the fee payable to GPs for notifications was ended in 2010.

Official numbers of whooping cough cases year by year therefore have to be interpreted very carefully to allow for the many factors that influence the accuracy of the reported incidence.

The solid brown line in Fig. 4.2 is the whooping cough vaccine acceptance rate which starts in 1967 at 78% and is more or less level until the sudden fall to 38% by 1976 after the scare that started in 1974. It is probable that the 1967 78% level went back to 1957. There is no good information on uptake between 1952, when it became available, and 1957 when it was officially recommended. The blue line is the number of notified whooping cough cases in England and Wales, which starts in 1940 when the disease first became notifiable. The rather big drop after 1941 is probably something to do with the 1939–45 war. There are peaks about every four years throughout, but there is an overall big drop starting in 1955 and continuing to diminish progressively to 1976.

On the face of it, the drop seems to coincide with the introduction of immunisation, but the eminent epidemiologist Professor Gordon Stewart of Glasgow believed the drop was a result of improved socio-economic conditions rather than immunisation and argued the case strongly to fellow academics and to the media, who gave his views considerable prominence.

It was well known amongst those giving triple vaccine that it sometimes caused worrying reactions and that these were more frequent than after diphtheria and

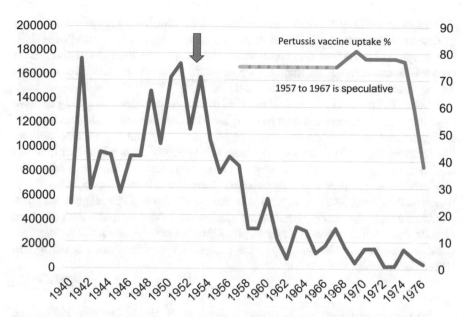

Fig. 4.2 Whooping cough notifications for England and Wales 1940 to 1976 (blue) and vaccine acceptance % from 1968 (brown). It first became available in 1952 (arrow) and officially advised in 1957. Uptake between 1957 and 1967 is speculative

tetanus alone. A few babies would become unwell within hours of it being given, with fever, irritability, refusing feeds, and sometimes inconsolable screaming that could persist for several hours. Often there was also inflammation at the injection site.

In my experience, and that of others doing the same job, these babies soon recovered, apparently without harm but we were careful to ensure that future injections were of diphtheria and tetanus alone. These reactions were a little worrying as they hinted there might be brain irritation of some sort going on.

My eldest daughter Nicola was severely affected by her first triple jab. She was feverish and screaming within hours of it. She spent 48 hours in Clatterbridge Hospital where I was then working as a senior house officer in 1969. Gladys Hemes, the conscientious paediatrician in charge, performed a lumbar puncture to exclude meningitis, such was her concern. Fortunately she made a full recovery and it didn't stop me giving my other children the triple vaccine when their turn came. They were fine but Nicola just continued the course with diphtheria and tetanus vaccine.[1]

---

[1] Nicola was 50 in January 2019 when I was writing this book. Her birthday celebration was cancelled because she was ill with a chest infection. She was off work for two weeks before struggling back. Then her 15-year-old son Thomas developed a cough with choking attacks that made me suspect whooping cough. Nicola was tested and pertussis was confirmed. Thomas clearly had it too, but he will not be counted in the 'laboratory confirmed' statistics.

There seemed little doubt that immunisation had caused a big drop in the number of cases and the number of deaths from whooping cough, so the occasional short illness caused by the injection was a small price to pay for lives saved and misery avoided. Because it was an impure vaccine made from whole, killed bacteria, we just accepted that some reactions would occur.

Doctors did not, and do not, find this kind of thing especially concerning because most medical treatments and interventions are a balance between risks and benefits, a conflict that most antivaxxers[2] understand well. They and the vaccine hesitant diverge from the mainstream over the next step which is measurement of that balance between risks and benefits, which in most cases can be done with precision. The process is complex and mathematical, therefore difficult for most people to understand in detail, so we have to take the conclusions of the experts on trust. Antivaxxers and the vaccine hesitant are sincere but have a problem with this kind of trust. Paradoxically many prefer to place their trust in places most people consider untrustworthy. The whole vaccine compliance problem boils down to: 'In whom do we trust?'

*****

The whooping cough vaccine controversy had started in October 1973 when paediatric neurologist Dr. John Wilson of Great Ormond Street presented a paper to the British Paediatric Association describing 36 children he had seen over 11 years who had developed neurological problems within a week of being given triple vaccine. It might not have caused concern outside the medical profession but for the intervention of Jack Ashley MP who was himself profoundly deaf and a champion of campaigns for the rights of the disabled. He had recently achieved considerable political success by securing compensation payments for children damaged by the drug thalidomide.[3] On 29th January 1974 he asked a series of questions in parliament about adverse effects following vaccination, including the possibility of a vaccine damage compensation scheme.

He was working closely with a pressure group called the Association of Parents of Vaccine-Damaged Children which had been set up originally to claim compensation for children damaged by polio vaccination in the 1960s. To be now joined by parents of children who claimed to be damaged by whooping cough vaccine gave strength of numbers and a greater moral justification for claiming compensation for individuals damaged by an action designed to help the whole of society. The perceived connection with thalidomide was strong, and Jack Ashley wanted to extend the government's compensation policy to innocent victims of vaccine damage. His action was well meaning and he could have had no idea that pertussis vaccine uptake would be affected so adversely.

---

[2] This does not apply to antivaxxers fundamentally opposed to the process. (On religious grounds for example.)

[3] Thalidomide had been a new drug in 1957 to treat pregnancy morning sickness but it caused devastating limb deformities.

Dr. John Wilson's Great Ormond Street team published its report [1] in a medical journal early in 1974 and this was followed in the April by an ITV programme on the subject, in which Dr. Wilson stated his belief that the vaccine was linked to brain injury. Some months later on 25th September 1974, a *Daily Mail* editorial stated:

> ....we are not told that 80 children every year suffer brain damage because of whooping cough vaccine, and just as many die. We are not told that there are possibly hundreds of teenagers with the body of an adult and the mind of a child because they were vaccinated.

This was a gross distortion of the known facts at the time and was unsupportable, but proving it wrong was difficult. Most doctors I knew took the same view as I did, that the benefits outweighed the risks, and therefore we continued to recommend it. Brain damage cases, if they occurred, were believed to be very rare, but without immunisation we believed hundreds of babies would die every year.

There were other doctors who took a different view. They stopped using the triple and reverted to diphtheria and tetanus alone for all their patients. Many of those who still used the triple often played safe by avoiding it in any child who had a history of any condition they thought might increase the risk of a reaction, such as a difficult birth or allergy. This excessive (and, we now know, unjustified) caution tended to fuel parents' doubts, thus reinforcing a vicious circle of fear.

Most parents now brought their baby to the immunisation clinic having made a clear decision not to have the whooping cough vaccine, and it was impossible to persuade the attending parent otherwise. If both parents had been present, there might have been a chance of doing so but they hardly ever were. One parent would not change the decision alone. Within two years of the media taking up the cause against whooping cough vaccine, less than half of babies were being immunised against whooping cough in the UK.

The dispute was continuing amongst doctors in the medical journals. The most vocal academic opponent of the vaccine was Professor Gordon Stewart of Glasgow who was convinced, and he provided evidence, that the drop in case numbers was the result of improved socioeconomic conditions rather than the vaccine. He argued his case in *The Lancet* [2] and the newspapers. The pro-vaccine case was made by the official government body, The Joint Committee on Vaccination and Immunisation (*JCVI*), chaired by Professor Sir Charles Stuart-Harris who wrote in the *British Medical Journal, The Lancet's* traditional rival. There were many other experts who joined in the dispute. The problem was that there was no firm evidence to prove the case either way.

In response to the controversy the government asked *JCVI* to quickly produce a report setting out the known facts and to make recommendations about the use of the vaccine. It was published in May 1977 [3]. It recommended continued use of the vaccine as it believed it was effective. It accepted that there was uncertainty about whether the vaccine caused permanent harm occasionally and explained that they hoped to answer the question definitively by setting up a special study. This study would analyse every case of brain inflammation admitted to hospital over the next few years and check whether there was a relationship with previous whooping cough immunisation. It was called the 'National Childhood Encephalopathy Study' and was expected to answer the question of safety once and for all, but in the event it took four years until 1981 to gather enough data to come to any conclusion.

Things could hardly have been worse for vaccine acceptance. The rate had fallen drastically, and the government, through the Department of Health, was well aware that the situation would almost certainly lead to a big outbreak of whooping cough in 1978 and would inevitably include a large number of deaths in babies.

With regard to compensation there was publicly great sympathy for a concept which seemed only fair, but the idea unfortunately clashed with the government's policy to improve social and financial provision for *all* disabilities, not just specific groups. It therefore resisted Jack Ashley's calls and wanted to wait for a scientific answer to the issue of whether it was truly damaging, so put off the compensation question by the delaying tactic of referring it to a legal team.

Eventually after three years in 1979, it was considered politically expedient to agree to interim payments of £10,000 where there was circumstantial evidence of damage associated with administration of vaccine, but cases were not fully settled until many years later. There were about 700 payments in all.[4] Anecdotal reports circulating at the time, from members of the assessment panels, suggested that the impossibility of disproving a connection in many cases was the reason for granting payments simply out of sympathy.

Many people now believe that the failure to agree early compensation fuelled the controversy adversely and prolonged it by maintaining it as a worthy cause. The government saw paying compensation as acceptance that the vaccine caused damage, for which there was dubious evidence and would therefore be counterproductive in the struggle to raise the immunisation rate.

It would be almost a decade from the start of the controversy in 1974 before the answer to the two big questions could be answered with confidence. Does the vaccine work? Does it cause brain damage?

References

1. Kulenkampff M, Schwartzman JS, Wilson J. Neurological complications of pertussis inoculation. Arch Dis Child. 1974 Jan 1;49(1):46–9.
2. Bassili WR, Stewart GT. Epidemiological evaluation of immunisation and other factors in the control of whooping-cough. Lancet. 1976 Feb 28;307(7957):471–4.
3. Whooping cough vaccination. Review of the evidence by the joint committee on vaccination and immunisation. HM Stationery Office. 1977.

---

[4] Subsequent research showed that the damage was a consequence of fever and underlying undiagnosed epilepsy and would have occurred with any feverish event. The vaccine reaction was the trigger, not the cause (see Chap. 7).

Chapter 5
1977–9. The First Outbreak of Whooping Cough

The very first cases of whooping cough in Keyworth started in August 1977, but I was unaware of them until several weeks had passed. Even then they were regarded as a bit of a curiosity and a few more weeks went by before we realised we were being overtaken by a most unusual event, an outbreak of whooping cough of seemingly epidemic proportions.

Keyworth Health Centre provided services for a much wider area than the village itself which had a population of about 8000. There was Tollerton, about two miles to the north with another 2000, and scattered east, west and south were the many smaller villages and hamlets we looked after: Plumtree, Bradmore, Bunny, Wysall, Widmerpool, Clipston-on-the-Wolds, Normanton-on-the-Wolds, Stanton-on-the-Wolds, Willoughby-on-the-Wolds, Hickling, Hickling Pastures and Kinoulton. The total practice population at that time was 11,800 and differed somewhat from other places insofar as almost all children attended a nursery or a pre-school playgroup regularly until school age, making the spread of infectious disease more likely; a factor that I believe played a significant part in this particular community.

Classical whooping cough as described in textbooks is said to have three phases, a catarrhal phase with a runny nose, sore throat and a bit of a dry cough and a mild fever that lasts a week to 10 days, followed by a paroxysmal phase of terrible coughing, vomiting and whooping for two to four weeks, then a convalescent phase in which the symptoms slowly get better over two to eight weeks. There is a problem with classical descriptions as any doctor will tell you. In reality very few real patients present with classical symptoms of anything. They present with one or perhaps two of the symptoms, and the doctor has to fill in the blanks to arrive at a diagnosis, which might be one of a dozen possibilities. That is when experience, knowledge and skill are applied. It is difficult in real life to go from a patient's symptoms to a diagnosis, and whooping cough is a good illustration of where that is the case.

Whooping cough can last for a long time. Three months is not unusual. In China it is called the hundred-day cough. It can also be quite mild in the sense that the person affected may have no bodily disturbance at all, that is to say, no fever, pain,

© The Editor(s) (if applicable) and The Author(s), under exclusive license to
Springer Nature Switzerland AG 2020
D. Jenkinson, *Outbreak in the Village*, Springer Biographies,
https://doi.org/10.1007/978-3-030-45485-2_5

or loss of appetite; just a choking attack of coughing a few times a day, maybe just at night, otherwise carrying on as normal. So many, if not the majority of parents of a child with symptoms like this, may not be worried about it. After all, children are always getting coughs and colds, they no doubt think.

It was my senior partner Manson Russell who first mentioned that he had seen some children with whooping cough. It was just a passing comment and he did not attach any unusual significance to it. Manson qualified in medicine at the Middlesex Hospital in 1947 and started in general practice in Keyworth in 1948, when seeing scores of cases every few years was normal. Diagnosing it from the sound of the cough or the story given by the parents was second nature to him and hardly worthy of comment. I think he wanted me to know because of my interest in paediatrics. He was possibly unaware of the impact of the vaccine controversy and the expectation of a national outbreak, whose start he was one of the first to witness.

My ears pricked up at this point as I asked who they were because I knew I needed to inform the health visitors. Gwen had now been joined by a new health visitor, Jackie Pepper, who remained at Keyworth for many years and assisted me constantly and directly with my whooping cough studies. I also knew that the cases needed to be notified to the local Medical Officer of Health, and that Manson, like so many crusty, salt-of-the-earth GPs in those days, probably would not bother with that form-filling.

Something made me start keeping notes of the whooping cough cases at an early stage. In those days, long before computers, there was no easy way of retrieving information from the standard medical record envelopes without repeatedly going back to them, so a separate record was required. I used an exercise book and started a list. I had my course project partly in mind and I also felt curious to find out for my own satisfaction, whether being immunised was beneficial, and this was possibly a way to a quick answer.

It was mid-October before I saw that cases were increasing rapidly and there was probably going to be a large number of children affected. A month earlier in September, when there were few cases, I found I could work backwards and discover where an individual had caught it, but as the numbers escalated it became more difficult.

I wrote normal records as I saw new cases and also put them in my book, according to the date of presentation, so consequently they were not in chronological start-date order. This is because some presented within a week of starting symptoms, while some were discovered almost by accident up to three months later. After a few months the limitations of a book were becoming apparent as I realised cases were probably just going to keep coming, so I changed to an A4 pro-forma I designed for the purpose and filed them in a ring binder.

*****

To properly describe the progress of the outbreak I need to describe individuals, but I cannot use real names because of the need for confidentiality.[1] Patients must

---

[1] Pseudonyms would be easier but for every pseudonym there would be a real patient with that name!

not be able to be recognised by others. I have devised a system to get around this using the age, sex and case number. First comes the age, 1m to 99y, second the sex, 'm' or 'f', and then the case number, 1 to 744 in a square bracket. So for example, the first sheet in my file is 3yf[1] ('three-year-old female 1') and the last is 64f[744] ('sixty-four-year-old female 744').

The very first case was 3yf[14] who started with a cough on 6th August 1977. Let's call this family the 'Jones' family. Her baby brother 5mm[15] started coughing on 13th August. It was not known from whom 3yf[14] had picked it up but the family had been on holiday on the Isle of Wight two weeks previously, so it was assumed to be from somebody there in Hampshire. Neither of them had been immunised against whooping cough. There was also a seven-year-old immunised daughter in the family who did not catch whooping cough. She was, however, in the same school as 5yf[3] who started coughing in mid-August but without a known source for her infection. In the above instance it is tempting to speculate that the older sister who was immunised acquired an asymptomatic or subclinical *Bordetella pertussis* infection and passed it to a schoolmate.

This 'Jones' family lived in a small cul-de-sac at house number one. Three weeks later the two boys at number two developed it. They took it into one of the three primary schools in Keyworth.

4yf[22] started on 20th August. I don't know where she caught it, but she lived on a road at number 20. Her sister 17mf[21] also had it. Between them they spread it to 3yf[8] at number 26 and her playgroup, also to 3ym[11] and 14mf[12] at number 22, also to 2ym[63], and thence to his big brother 4ym[62] and on into yet another playgroup.

Thereafter the numbers increased so much the transmission route was confused by too many alternative pathway possibilities, but in every outbreak of ensuing decades this kind of pattern was usually clear in the early stages. It is the pattern one would expect to see if individuals who are susceptible get recognisable symptoms of the disease and if one of those symptoms is responsible for transmission. It does not support the idea that infection can be passed on readily by somebody without symptoms. Nevertheless, there were some instances where it was tempting to explain how it was caught in that way, and we knew it was theoretically possible.

Some infectious diseases could only be diagnosed clinically at that time (1977), that is to say, purely from the symptoms, without any confirmatory laboratory test; mumps and measles for instance. But for others, such as typhoid, scarlet fever and tuberculosis, it was possible and necessary to arrange for the laboratory to culture the micro-organism responsible to confirm a diagnosis. All micro-organisms are different and require different methods of culture. *Bordetella pertussis*, the cause of whooping cough, is not only difficult to grow, but it is also difficult to find because it inhabits the back of the nose and the lung passages, and those places are not easy to get at. It is common when looking for certain types of micro-organism that cause disease in the respiratory tract (nose to lung) to take a swab from the nose or throat or collect a specimen of sputum. That can work very well for many infections, but it does not work for culturing *Bordetella pertussis*.

There was an old-fashioned method using a 'cough plate' that was popular in the early twentieth century, and when done properly was capable of achieving the best results. A petri dish with culture medium in it had to be held in front of the mouth as the patient had a paroxysm of coughing so that droplets which might contain *Bordetella pertussis* would land on the culture medium within the dish. This could be done in hospital easily enough but at home or in general practice quite impossible, as you could not sit around for a couple of hours waiting for a paroxysm to happen (although I have seen some old-time doctors write that they had foolproof ways of inducing a paroxysm).

Another method was to take a swab from the back of the nose. Such per-nasal swabs are specially made for the purpose. They have a tiny fibrous Dacron[2] tip on the end of a flexible braided steel wire about 12 cm long (Fig. 5.1). To collect a specimen the swab has to pass through a nostril horizontally (if the patient is standing) until it is stopped by the tissue at the back of the nasopharynx. It is then withdrawn. It is a very uncomfortable procedure to have done. Even when done quickly and skilfully the patient usually recoils with an expletive, eyes watering and possibly sneezing into the bargain. Very few adults will permit it to be repeated, and it can put a child off visiting a doctor for a long time. This reason alone was enough for doctors to avoid it in general practice, but there was another difficulty. The microorganism concerned is very delicate and fastidious in its growth requirement. Organisms on any swab taken in general practice would be dead and therefore undetectable by the time the swab arrived at the laboratory many hours later.

I could not go on diagnosing whooping cough at the rate I was without attempting to confirm it bacteriologically, so I set off to talk to my local laboratory at Nottingham Queens Medical Centre. I met one of the consultant pathologists in the microbiology department, Dr. Purvin Ispahani. The junior staff all seemed to be terrified of her and couldn't understand why I was so keen to see her. She was small, had a loud commanding voice and clearly stood no nonsense. She was not only extremely helpful to me but went out of her way to ensure that the arrangements we made were carried out to the letter by her staff.

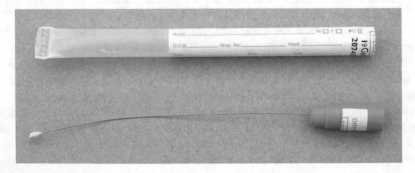

Fig. 5.1 A per-nasal swab for detecting *Bordetella pertussis* alongside its container

---

[2] Cotton inhibits the growth of *Bordetella pertussis*.

I explained that I had already diagnosed 38 cases clinically and needed to get bacteriological confirmation. She must have quickly worked out a plan in her head. It was brilliant and just perfect for the situation. I would take per-nasal swabs just from cases in the early stages of the disease, because the children would already have been coughing for a week or so before being suspected as whooping cough, which did not leave much time to find the bacteria, which are usually undetectable after three weeks. But what was I going to do with the swab after it was taken, to keep the bacteria alive?

She brought out some petri dishes that were used for culturing bacteria, already coated on the bottom inside with the appropriate culture medium. I would need to inoculate a corner of the agar medium with my swab and rub the remainder of the agar with the swab in a zig-zag pattern that maximised the chance of being able to identify individual colonies if they grew. I would mark each dish with the name and date of birth of the patient and also fill in a laboratory form separately with full details.

There was yet more that needed doing if it was going to work. *Bordetella pertussis* likes an atmosphere rich in carbon dioxide to grow. Dr. Ispahani produced a tall cylindrical tin box about six inches in diameter with a tightly fitting lid into which a dozen petri dishes could be stacked. Each day after I had collected all my samples and plated them out, I would need to stack them in the tin, place a burning candle stub on top and put the lid back on. The candle would soon be extinguished from lack of oxygen but would have generated the carbon dioxide-rich atmosphere required. She then instructed one of her laboratory technicians who lived in Keyworth to call into the Health Centre every morning to collect the previous day's plates. Her ability to command and control was awesome.

I couldn't wait to get started. The first plating out was made on November 14th and I had done 12 by the end of 17th. I eagerly awaited the results knowing it took at least five days for *Bordetella pertussis* to grow. About a week later in the middle of a morning surgery the telephone rang.

'Dr. Ispahani from the Public Health Laboratory for you'.

'Good morning Dr. Ispahani'.

'Good morning Dr. Jenkinson. I would like to congratulate you on your diagnosis. Eight of the twelve have grown *Bordetella pertussis*'.

I was amazed. I was not expecting such a good result (Fig. 5.2). It gave me the boost I needed to continue what I was doing.

Later I learnt of her commitment to the project. It can be very difficult to interpret the growths on microbiological cultures. You need a lot of experience, care and knowledge to spot what might only be a single colony with the right characteristics among a multitude of different colonies. She scrutinised every plate personally to ensure nothing was missed. That was the way she worked, and I am so grateful to her. She also sent off every isolation, as was required, to Dr. Noel Preston at the Pertussis Reference Laboratory in Manchester, where precise details of the currently circulating strains were determined (to the extent possible in those days), and the intelligence used to inform epidemiologists and vaccine manufacturers.

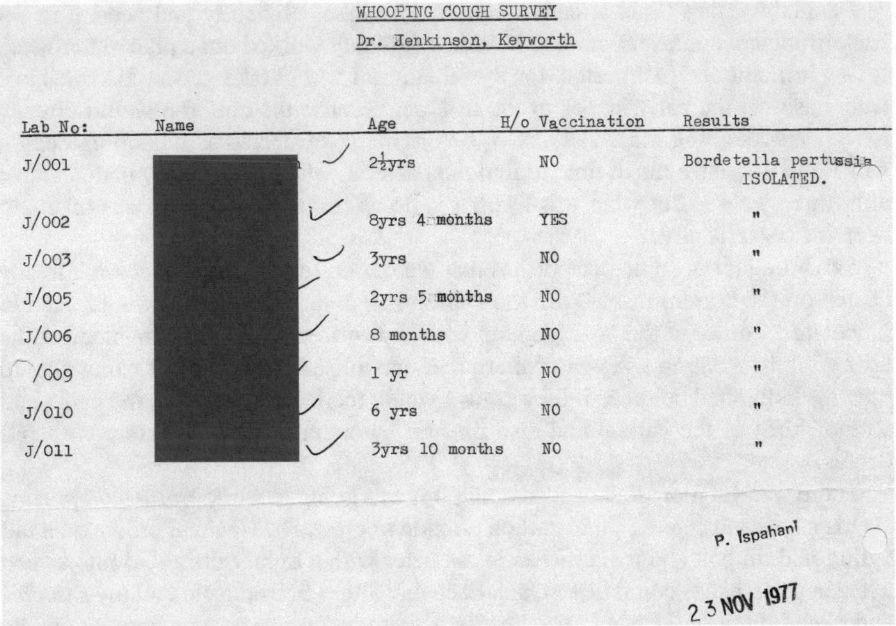

WHOOPING COUGH SURVEY
Dr. Jenkinson, Keyworth

Lab No:	Name	Age	H/o Vaccination	Results
J/001		2½yrs	NO	Bordetella pertussis ISOLATED.
J/002		8yrs 4 months	YES	"
J/003		3yrs	NO	"
J/005		2yrs 5 months	NO	"
J/006		8 months	NO	"
/009		1 yr	NO	"
J/010		6 yrs	NO	"
J/011		3yrs 10 months	NO	"

P. Ispahani

2 3 NOV 1977

Fig. 5.2 The first lab report of the 8 positive cultures from 12

Eight out of twelve was a higher proportion than expected or that I subsequently ever achieved. It might have been luck or more likely to do with the fluke use of the amoxicillin antibiotic. In the early stages of the outbreak many children were given this for their symptoms before it was realised it might be whooping cough and those patients were some of the ones I was taking swabs from. Amoxicillin has no effect on *Bordetella pertussis,* but it does kill many other bacteria that live in the nose. It may therefore have made *Bordetella pertussis* isolation more likely, as there would have been less contamination of the swab with these other bacteria that could swamp the ones we were looking for. Soon after, we all changed to using erythromycin, the correct antibiotic, which rapidly kills *Bordetella pertussis*.

After a slowish start the epidemic really took off. I saw seven cases that started in August 1977 and eight in September. October saw the start of 27 cases; November, 74; December, 46; January, 17; February, 13; and March, 5 (Fig. 5.3); making a total of 197 in eight months.

At the height of it in November and December we were seeing two or three new cases every day. It was not as overwhelming as might be expected. There were four doctors seeing about 40 patients each, every day. The whooping cough cases were in numerical terms insignificant compared with the main bulk of our work. There was slightly more work for me because I had taken an interest, and cases were directed to me if possible, sometimes by receptionists where it would become part of the normal workload through the appointment system, or sometimes directly from a partner in the next room who might want me to take a swab from a new case.

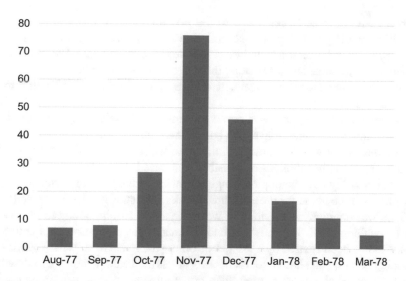

Fig. 5.3 Monthly totals of whooping cough cases in Keyworth

Health visitors who had found a case might push it onto me too. There was also the extra paperwork I had to do, but by and large it did not change our day to day running.

Schools and nurseries were affected to a much greater extent. Once it became clear there were cases of whooping cough, they were very keen to keep cases out and anyone with a cough was advised to get it checked out with a doctor. For many months their attendances were seriously down as a consequence. Unfortunately it was not easy to tell who had it and who hadn't. This was wintertime when coughs and colds, especially among children, are very common. Whooping cough starts just like many other coughs and colds and only after a week or so develops into a choking cough with vomiting and whooping, making it impossible to tell in the early stages whether it was one or the other. There are no signs that a doctor can detect on examination to distinguish them, and it is most infectious in the early stages and easily passed to susceptibles. That is why it spread so explosively in Keyworth and surrounds because most children attended a nursery or playgroup or school. Even taking a per-nasal swab was no help because it took five to seven days to grow the bacteria. If everyone with a cough or cold had been excluded, there would have been almost nobody there!

There was a rule book to give guidance to schools and nurseries about infectious diseases. In general it advised to exclude whooping cough cases until recovered. This advice had been written decades before when whooping cough was viewed very seriously. There had been no real need to update it because everyone thought it had for practical purposes disappeared. A more enlightened approach was taken by some head teachers and nursery leaders who had sought out more up to date advice, which was based on the fact that sufferers would not be able to pass it on after three weeks and could be allowed back, although in practice most are at their worst symptom-wise at three weeks and it would be impossible to readmit a child who was still coughing and vomiting several times a day. We soon after learnt that *Bordetella pertussis* bacteria were best eradicated by erythromycin. We usually prescribed a 7-day course and understood that after 3 to 5 days the person was no longer infectious and could mix again. It took a considerable while for this to be accepted by everyone.

The general effect of this was confusion amongst parents, who really did not know anything about whooping cough but believed it to be a severe illness. Most of the children with it were not unwell in themselves and only coughed occasionally, so their parents either thought or possibly convinced themselves that their child could not possibly have whooping cough, knowing in the back of their minds that he or she could be home for the next three weeks at least, if it were confirmed!

There was definitely *not* a flood of worried parents rushing along to the doctor with a coughing child. The most common scenario was a parent with a coughing child who had been spotted by a teacher or other parent and told, 'That sounds like whooping cough, you should get it checked out at the health centre', and they had reluctantly come along. More often than not it turned out to be whooping cough, because it is very easy to diagnose once you have heard somebody with it, and I think the whole village became good at it. I will try to explain why this is.

Think for a moment what happens when you feel you have to cough. You take a breath in and force it out in three or four short bursts making a coughing noise in the process and then you breathe in again. If you still feel that tickle in your upper chest you repeat the process until the feeling has gone. It might come back ten minutes later, and you repeat the whole process. We might represent it thus:

Inspire—cough—cough—cough—inspire—cough—cough—cough—inspire—cough—cough—inspire.

In typical whooping cough you lose the ability to breathe in until your lungs are empty of air and even then there may be a delay of many seconds before you can do it. Like this:

Cough—cough—cough—cough—cough—cough—cough—cough—cough—cough—cough—cough—(red face then blue face then vomit)—inspire (possibly repeat whole process immediately).

After you have recovered from the above you may well be absolutely fine for a couple of hours or longer. This kind of cough is known as a paroxysmal cough. In common parlance a choking cough, albeit a very severe one. Several paroxysms may follow one after the other. The inability to breathe in again after a paroxysm is variable according to the severity. It is a very frightening experience and is identical to a feeling of suffocation. If it goes on for 20 seconds or so, which it sometimes can, the patient will go blue (cyanosis is the medical term). Older children and adults can usually tolerate this, although it is distressing in the extreme. Babies often cannot. If babies are also feeble or exhausted they may not start breathing again without resuscitation. These episodes of inability to breathe are called apnoea attacks. After a paroxysm it is common to vomit reflexly or salivate profusely or both. At the end of a paroxysm, during the eventual gasping inspiration a stridulous 'whooping' noise may occur.

Some people can faint after a paroxysm and may get muscular twitching, resembling a fit, which it is not. They, as in any faint, will usually come around quickly but it can be dangerous, because falling frequently results in injury, especially in a bathroom where one is surrounded by hard objects. This scenario is common but usually confined to adults, who often hasten to the bathroom when a paroxysm comes on.

*****

I needed to decide what criteria I would use to make a clinical diagnosis and stick to it if my statistics were to be of any value. It became obvious to me fairly quickly that the cases I was seeing were invariably lasting at least three weeks, so I made three weeks of paroxysmal coughing my criterion for clinical diagnosis. I did not invent this, many others had used the same; I merely confirmed for myself this was practicable. I did not need vomiting or whooping or apnoea in addition although these were all symptoms I recorded for each patient along with several other pieces of information about contacts and family members.

I stuck with three weeks for the whole duration of my researching. There were many occasions of patients with almost certain whooping cough who had less than three weeks of paroxysmal coughing. They were never included in the figures except on the rare occasions I obtained laboratory confirmation. These borderline cases were filed separately. There were 36 by the end of the study.

Everyone with possible whooping cough needed to be seen and carefully examined after taking a history. It is often relatively mild, but it can be severe and with complications, so I needed to know what severity of case I was dealing with, and if necessary, arrange follow-up checks. I was very unsure at this stage whether treatment was beneficial or not. It was usual to prescribe some kind of cough linctus in those days, either the expectorant sort to help bring up sputum or the suppressant sort to stop an irritating cough. The latter seemed more logical. They usually contained active ingredients such as an antihistamine or codeine and it certainly made the parents feel they were doing something to relieve the distress, but we never felt it made a contribution to recovery, although there was always a bit of positive feedback from parents, no doubt from their placebo effect.[3] We routinely prescribed the antibiotic erythromycin in the early stages to kill the bacteria in the hope of limiting spread. We were gradually coming to realise the disease just had to take its natural course. Some people could accept this and others not, so the latter expected us to do more than we were able. We tried to deal with this by providing close follow-up instead of a useless prescription.

My general impression after seeing a hundred or so cases was that only about half the children with whooping cough actually whooped, although many more had vomiting associated with it. Apnoea (temporarily stopping breathing) was much less common and seemed associated with more severe illness. The duration was

---

[3] Scientific studies later confirmed that standard cough remedies and various inhalers used in lung conditions have no effect on whooping cough.

more difficult to know because it was many months for some. In general it took about two months to clear up, although many were better after one month. I also recorded the estimated number of paroxysms per 24 hours. The maximum seemed to be about 24 or one every hour while some had as few as five. Most seemed to have between 7 and 12. In general the more severe it was, the more paroxysms there were and a greater likelihood of apnoea after a paroxysm.

Health visitors Gwen and Jackie spent a lot of time case-finding. They were able to get into playgroups and nurseries and talk to the leaders about who was coughing suspiciously. Without their detective work a great many of the cases would not have come to light. They were extremely good at it right from the start and got even better in successive outbreaks.

One of the problems for doctors who are checking for whooping cough is that paroxysms may only happen every couple of hours on average and sometimes only at night. So it is very probable that a child brought to the doctor to be checked for whooping cough will not have the slightest cough while there, even if they have to wait an hour to be seen! There was many a time when I would be told by a receptionist or health visitor that somebody in the waiting room had a cough that sounded like whooping cough, usually when I was running late, but by the time they saw me they were absolutely fine with no sign of illness at all. But the story of a paroxysmal cough was all I needed to take a swab, if I was allowed! I think by Christmas 1977 everyone in the village knew to keep away from Dr. J (as I was known) because 'He will try to take one of those horrible nose swabs from your poor child!'

This was one of the reasons for having less swab information than I would have liked. A baby or toddler could be held firmly by its mother as I took a swab, and at a young age it was definitely in the child's best interest to do so. Beyond the toddler stage it could become a serious contest if the child understood what you were going to try to do as I approached its nose with a long wire. At that age a swab was much less justifiable clinically, as they were old enough to get over whooping cough without significant worry about coming to harm, so I would back off from the procedure, much to the relief of everyone. Older children could be reasoned with and sometimes coerced or bribed even, but at best, less than half the patients would submit to it, and I fully sympathised with that.

I was really impressed with the way parents coped with several weeks of this illness, sometimes months. They were usually up at night several times to comfort their child's, or often several children's coughing attacks with vomiting and general distress. They just got on with it. I suppose there was no choice. It was perhaps harder for the ones who had their child immunised and still caught it. Strangely I don't recall any parent of an unimmunised child saying they wished they had had it.

I don't think I was expecting to see an adult with it, but 34ym[18] came to see me wondering if he could possibly have it because his cough seemed so similar to his daughter's (21mf[53]) cough, and she had had whooping cough for several weeks. She had been swab-positive. There seemed to be a general perception that it was only a disease of childhood and I did not know much better myself at the time.[4] There was a good reason for thinking this because it had been true for generations that children seemed to inevitably get a succession of diseases in childhood, often associated with fever and rashes, and it was a list that could be ticked off one by one, knowing that if your child got over it, they would not be attacked again by the same disease. Included were measles, mumps, rubella, scarlet fever and chicken pox. Some societies numbered them, the first being measles, known as 'first disease', and so on. We still see the hangover of this today in 'fifth disease', also known as 'slapped cheek syndrome', which is caused by a parvovirus and often causes mini-epidemics. Whooping cough does not cause a rash or much of a fever so was never numbered with the others but in folk health-lore it was often thought of as one of these childhood illnesses.

This unfortunate parent was getting about seven choking attacks of coughing followed by vomiting throughout the day and he had started about three weeks after his baby daughter started coughing. She had not been immunised and he was too old at 34 to have been. There could be no doubt he had whooping cough. This made me sit up, metaphorically, as I was exactly the same age as he was, and realised I must be at risk myself coming into contact with it so much. Fortunately, I never did get it, in spite of all the contact. I must have had good immunity acquired by some means or other, and possibly boosted by the repeated contact. I have no idea whether I had had it as a child. The aforementioned family also had an unimmunised three-year-old daughter who coughed mildly for about four weeks, but it was never paroxysmal. It probably was subclinical pertussis though!

I had not seen any serious complications in this outbreak. One child had been admitted to hospital because of a prolonged period of not breathing and she went very blue, frightening her parents and making me very nervous about what might happen, as they said it lasted two minutes. She was three years old but was discharged from hospital the next day.

There were lots of minor complications such as secondary bronchitis and middle ear infections, often requiring antibiotics and usually clearing quickly. Textbook descriptions describe a multitude of possible complications as a result of stressful coughing, but I have seen none of them. They include torn tongue frenulum, hernias and subconjunctival haemorrhage. Readers that know about whooping cough will be amazed that I have not yet seen a subconjunctival haemorrhage in any of my patients, but it is true.

One set of five-year-old identical twins had *Bordetella parapertussis*, a less frequent cause of whooping cough and one of the three[5] *Bordetella* species known to affect humans. It is usually less severe than *Bordetella pertussis*. Only one of the twins whooped and she also had more frequent coughing. One might have expected them to be identically affected but there must have been something else causing the difference. One of the very first cases in Keyworth also had *Bordetella parapertus-*

---

[4] Of 197 cases in this first outbreak, three were in adults.

[5] The third is *Bordetella holmesii.*

sis. His father was a dermatologist who diagnosed his son with whooping cough himself and had a swab taken to confirm it.

I tried to find out if what was happening in Keyworth was happening elsewhere in the locality. Dr. Ispahani said she was getting a few isolations from other areas but GP friends in the neighbouring town of West Bridgford were apparently not diagnosing it. To this day I have no idea whether it missed West Bridgford or whether my colleagues missed diagnosing it. The discrepancy can be partly explained by that fact that the outbreak peaked in November 1977 in Keyworth, while in England and Wales as a whole, the peak came later in 1978. West Bridgford had possibly just not been hit when I was inquiring. The figures for England and Wales however were definitely rising, with 1467 notifications in October 1977; 3128 in November and 6127 in December.

By about April 1978 it looked as if the peak for us in Keyworth was well over and I could start gathering the statistics together. I had not gained any particular feeling that the vaccine worked so had not done any preliminary work. Indeed, the overwhelming sense was that of disappointment that so many cases were in immunised children. I also knew that the maths could not be completed until I knew the numbers that were *unaffected*. When it all seemed to have settled, I started the work of analysis.

Chapter 6
Does the Vaccine Work?

Keyworth is a nice enough place but in 1977 the only reason a stranger might make a detour to it would be to visit the high standing fourteenth-century parish church with its lantern tower, where a fire used to be lit to guide travellers; apart, that is, from student teachers, who were attending the Mary Ward teacher training college in Keyworth. It had been opened in 1968 by Princess Anne and was a Roman Catholic institution operated by the Loreto Sisters under the administration of the University of Nottingham. Most of the students were young ladies from Ireland and many subsequently married and settled locally. Such was the demand for contraceptive services that we held a special clinic on their premises regularly. The availability of the contraceptive pill on the NHS was fairly new, and doctors were very cautious about prescribing it without a thorough medical check first. The partner slightly senior to me, Erl Annesley, had been saddled with this particular chore.

The irony of the place was its unfortunate address. It was on Nicker Hill, Keyworth. A rather posh address as it happens, but when the college opened, the local and national press thought it was a great joke. 'Nicker' happens to be a local term for a woodpecker, a fact that makes the joke more effective, or less effective, according to how far your imagination can stretch.

The college had but a short life and closed in 1977. The building is now much expanded and the headquarters of the British Geological Survey. Its current importance is somewhat greater than before and helps to keep Keyworth on the map, as it were.

*****

Although I had never done vaccine calculations before, the working out of whether an immunisation was effective or not seemed just a matter of simple arithmetic, so I dived in enthusiastically, truly not knowing what the result was going to be or even thinking for a moment it might be important. It was for my own satisfaction and also to have something to present to my course supervisor when the time came.

First, I worked out how many had been affected and unaffected in each birth year from 1974 to 1977, to give me the attack rate in the under-fives. These age groups were the crucial ones because in each birth year cohort there was a group who were immunised and a group who were not. I needed the numbers in both to work out the vaccine protection. Older children born pre-1974 were all immunised (before the scare), so there was no unimmunised group of the same age to compare them with.

There were 189 cases by the end of February 1978 and 126 of them were born in 1974 or later. They were all four years old or younger. The worst affected group was those born in 1974, about three years old, of whom one quarter had it. In the whole under 5 group, almost one in five caught whooping cough (Table 6.1).

Table 6.1 Whooping cough attack rates by birth year

Year of birth	1973	1974	1975	1976	1977	Overall
Attack rate %	18.6	25.1	18.1	17	10.3	18.7

To calculate vaccine protection each birth year cohort is divided into four groups, affected and immunised, affected and unimmunised, unaffected and immunised, and unaffected and unimmunised. This is seen in Table 6.2, taking just one year, 1973, for the sake of clarity.

Table 6.2 Vaccine protection for birth year 1973

Birth year	1973	
Immunisation	Yes	No
Number affected by whooping cough	9	17
Number unaffected by whooping cough	96	18
Vaccine protection	82%	

In the immunised group only 9 out of the 105 caught it but in the unimmunised group just about half of them were affected, 17 out of 35.

There is a standard way of calculating the vaccine protection and it uses the attack rates in the immunised and unimmunised. In the example in Table 6.2 the attack rate in the immunised is $9 \div (9 + 96)$, i.e. $9 \div 105$, which is 0.086 or 8.6%. The attack rate in the unimmunised is $17 \div (17 + 18)$, i.e. $17 \div 35$, which is 0.49 or 49%.

The vaccine protection is attack rate in the unimmunised minus the attack rate in the immunised, divided by the attack rate in the unimmunised. i.e. $49–8.6 \div 49$ which is 0.82 or 82%. The numbers for 1974, 1975 and 1976 are shown in Table 6.3.

Table 6.3 Vaccine protection by year of birth 1974–76

Birth year	1974		1975		1976	
Immunisation	Yes	No	Yes	No	Yes	No
Affected	7	35	2	23	0	24
Unaffected	84	41	74	39	43	74
Protection	85%		93%		100%	

The calculation cannot be done on the 1977 cohort because immunisation was not completed until 11 months of age so there was no immunised group to be compared with, although 9 were affected out of 87. If all four years with data are added together the vaccine protection was 84%.

When the same calculation was applied to siblings under six years of age in home contact with whooping cough the result came out at 74%, quite close to the 80% that was required before the first vaccines could be licenced (Table 6.4).

Table 6.4 Home contacts under six years

	Immunised	Unimmunised
Affected	5	26
Unaffected	21	7
Protection	74%	

I was able to compare the severity in the immunised and unimmunised groups, but I was limited by the fact that one of the measures is duration of coughing. Unfortunately for the analysis, many children had not stopped coughing at that time, so it could not be done. I did however find that apnoea and cyanotic attacks were much more common in the unimmunised group.

I was slightly disappointed with the result in a way, as I thought that all I had done was confirm what most people believed anyway. If I had shown it didn't work at all, well that would have been something worthy of being published! I spent all my spare time for weeks on end analysing the data in as many different ways as I could, all manually of course in 1978, with numerous scribbles on sheets and sheets of paper and occasional use of a calculator. Having done all the work, I thought it would be good exercise to write it up as a scientific paper even though it merely confirmed what we already believed to be true, as that would be the format I would use to present it to my Community Child Health class. Margaret Smith our dedicated and uncomplaining practice secretary was kept busy typing and retyping my revisions.

I did numerous rewrites until I was happy with it and included tables similar to the ones on previous pages. I included a diagram representing how I thought it had spread from case to case in the initial weeks and pointed out that the disease seemed less severe in immunised children since they had fewer instances of cyanotic attacks than the unimmunised. The paper's conclusion was that the vaccine appeared as effective as it was supposed to be, and that when it failed to protect, the resulting disease was less severe. I had no sense whatsoever that my results were important. From my perspective I felt I was one of 30,000 other GPs in the country, seeing the same patients, providing the same medical care and any one of them could have done the same thing and presumably obtained the same result. I don't think I appreciated the extent to which Professor Gordon Stewart's propaganda[1] had been effective in spooking those in power.

---

[1] I use the word propaganda deliberately because he disparaged good evidence that was contrary to his entrenched view.

I gave a copy to Professor Hull's secretary with some hesitation and asked her to see if he would be kind enough to look at it. I was not expecting him to be much interested because at that time I was not familiar with all the medical literature on the controversy or even the historical background to some of the issues. I found that out later, along with the knowledge that *JCVI* was considering the possibility of stopping pertussis immunisation as Japan had already done.

David Hull must have looked at it quite quickly because his secretary rang back soon afterwards to say it was ready for collection and the Professor had suggested some changes. I was much encouraged by his attached note (Fig. 6.1) but a little disconcerted by the amount he had rewritten, but he had neatly converted a bit of a muddle into something much clearer and coherent.

David Hull sent a copy of the paper to the professor of Microbiology at Great Ormond Street, Alastair Dudgeon, whom he knew to be an expert with a finger on the pulse of pertussis issues. His reply was very positive, pointing out areas needing clarification and suggesting trying for publication in the *BMJ* or *The Lancet*. After being rewritten again, with the help of David Hull I posted it off to the *BMJ*.

In the days before super-specialisation the *British Medical Journal* was the most prestigious medical journal in the UK with worldwide standing and a very large readership, with *The Lancet* second. Nowadays it is equally prestigious but focused on a more general readership, with a paper version and an expanded electronic version. Specialist journals now cater for the important research findings in the multitude of specialities and subspecialties that have emerged. I was thrilled to get an acceptance note (Fig. 6.2).

That seemed to be the end of the matter. The paper had been accepted, the mini-epidemic was over, and my paediatric community health course was also over, having presented my findings to the class on the last day along with all the other students. I think everyone was relieved that the vaccine worked so we could continue promoting it to parents as we had already been doing, but with our fingers tightly crossed because of the still unresolved brain damage question.

It was good to see it in print at last (Fig. 6.3) [1]. They had put it into the Epidemiology section on 19th August 1978, almost as a news item, which in a way it was, being the first measure of pertussis vaccine effectiveness in the UK for 30 years. They had edited out a few sections and omitted my diagram of the early pattern of spread, but it was otherwise unchanged.

England and Wales recorded 9000 notifications of whooping cough in the following month of September 1978, the highest number seen for a quarter of a century.

Although the Keyworth outbreak had come and almost gone in six months, for the remainder of the country it was ongoing. The total number of cases for 1978 was 65,957 in England and Wales. It then fell to 30,816 in 1979 and to 21,131 in 1980 (Fig. 6.4). There had been 32 known deaths from whooping cough in the four years 1977 to 1980, although the true number was probably up to seven times greater. This was because of the difficulty of recognition and confirmation, which resulted in the complications of the disease being the only recognisable causes of death. Careful analysis of all baby deaths at that time strongly suggested that many cases

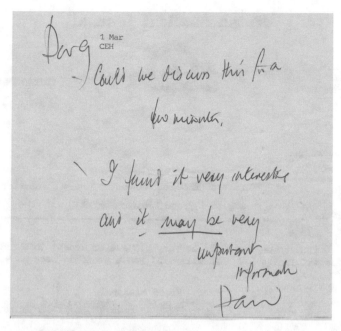

Fig. 6.1 'May be very important' note from David Hull

of death from bronchopneumonia [2] and sudden infant death syndrome [3] were actually due to *Bordetella pertussis*.

The 'Outbreak of whooping cough in general practice' paper was generally welcomed. I received several letters of congratulation from some eminent doctors including Sir Charles Stuart-Harris, chairman of the *JCVI*, who pointed out very politely that I had made an elementary error. I had used the attack rates in unimmunised and immunised as a ratio and called this vaccine protection. It was actually the odds ratio.[2] It only made a difference of a few percentage points and didn't change the overall result, but I felt like an idiot and learnt a valuable lesson. I had assumed that the clever people who reviewed and edited the material I had sent would spot an ignorant author's stupid errors. But no, it doesn't automatically happen. The author is completely responsible for everything. In most such instances the editor would probably see such an error as a good reason to reject the paper!

Pertussis vaccine arch-antagonist Professor Gordon Stewart made several criticisms that were published in a *BMJ* letter. He was sceptical about diagnostic accuracy and suspected different social conditions within my cases might have distorted the result. He also said that the attack rates were inconsistent with the older cases having been exposed to the disease in the 1974–5 outbreak. He did not take account

---

[2] The calculation of vaccine protection in previous pages is done correctly. Only the original paper has the error.

British Medical Journal

Telephone:
01-387 4499

Telegraphic Address:
AITIOLOGY LONDON W.C.1.

B.M.A. HOUSE
TAVISTOCK SQUARE
LONDON, WC1H 9JR

6 July 1978

Dear Dr Jenkinson

Paper entitled:
                An outbreak of whooping cough in general practice
                        August 1977 to February 1978

    Just a line to say we shall be very pleased to accept this article
for publication in the BMJ, and will send you a proof as soon as we can.

                    Yours sincerely

                    Stephen Lock MA MB FRCP
                    Editor

Fig. 6.2 Acceptance note from the *British Medical Journal* 1978

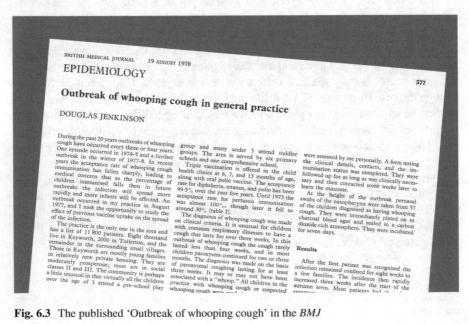

Fig. 6.3 The published 'Outbreak of whooping cough' in the *BMJ*

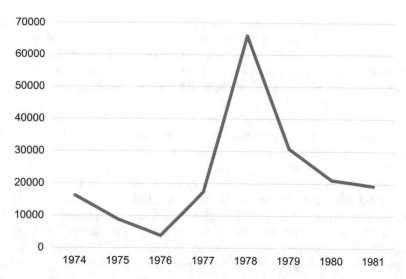

Fig. 6.4 Whooping cough notifications in England and Wales 1974 to 1981 showing the 1978 peak

of our very high vaccine uptake rate of 95% which probably prevented any cases occurring in those years. If there had been any I believe I would have seen them.

It being just a small study I was able to get the results out quickly, but several other researchers came out with reports soon afterwards. Yet others had embarked on larger and more elaborate investigations that necessarily took longer to conclude and publish. There was hope that these larger studies that were being organised by the Royal College of General Practitioners would yield firm information about the vaccine based on large numbers.

Two small studies similar to mine and one medium-sized one were published the following year in 1979 but we had to wait until 1981 for the two large ones. The first two small ones caused a lot of head scratching as they were both from the Scottish island of Shetland and had managed to reach completely opposite conclusions.

The first, in the 28th April 1979 *BMJ*, was from the chief administrative medical officer in Lerwick, Shetland, Dr. JD MacGregor, who analysed notifications of one- to four-year-olds with whooping cough. He knew the immunisation status of this population of 233 children, so he was able to calculate the vaccine effectiveness, which was 70% in boys and 80% in girls. So far so good!

The next report to come along, on 16th June 1979 in the *BMJ*, was initially a bit of a bombshell because it showed no difference in attack rates between the immunised and the unimmunised. The author, Robert Ditchburn, a GP also in Shetland, had experienced an outbreak at the same time as Keyworth and with similar numbers affected. He had stopped using whooping cough vaccine in 1974 altogether. A large proportion of his 5- to 16-year-olds were affected, almost all of them immunised and it was a similar proportion in his unimmunised under-fives. This could only be a valid comparison if the effectiveness of the immunisation was constant and long lasting. The two groups were not really comparable because of the massive age difference, but the editors must have thought it reasonable to publish any data which might shed light on the topic, and they would certainly have been anxious to maintain a balance. To not publish it at that time because of a possibly flawed method might look like prejudice. One could not help but note, however, from reading the acknowledgements at the end of the paper that Professor Gordon Stewart had contributed to its production.

The third paper of 1979 came out the next month in *The Lancet* and was an analysis of notified cases in Hertfordshire. One thousand and thirty-four cases were notified in the under-fives. The author showed that the vaccine effectiveness in notified cases was 94%.

Meantime I was still seeing a quite a few cases in Keyworth, 13 in 1979 but just five in 1980. We were still very conscious of the low immunisation rate and we needed to improve it. It was 31% in England and Wales for children born in 1978, the lowest it would ever be. The 1979 birth cohort still only reached 35% in spite of strong official endorsement, so we could see that the next outbreak which was due to start in 1981 (because outbreaks occur every four years) could also be big if not bigger.

A change that had come about in the way we dealt with per-nasal swabs meant that life was a bit simpler in that respect. The laboratory was now supplying transport medium for all bacterial swabs going to the laboratory and it was perfectly suitable for *Bordetella pertussis*. After taking the swab, instead of popping it back in the same dry tube or in our case, plating it out, it went into a similar tube partly filled with an agar gel transport medium that the organisms were happy to live in until the laboratory could deal with it further. This technique continued to serve us well for the next 20 years or so until blood testing for *Bordetella pertussis* replaced it.

January 1981 saw the publication of the first of the two big studies of whooping cough organised under the auspices of the Royal College of General Practitioners [4]. The Swansea Research Unit under the direction of Dr. WO Williams studied 2295 cases with the help of home visiting specialist nurses, identified directly or indirectly through the notification system. They studied the symptoms and complications as well as the effect of immunisation on those affected in the county of West Glamorgan (pop. 360,000). There was a lot of social deprivation in this area and the vaccine acceptance rate there had fallen to the lowest level in the UK, just 9.5% for 1974 births. There were three deaths and significant complications too. Pneumonia accounted for 18, 14 patients had convulsions and 64 were admitted to hospital for a variety of other reasons, many being social. The vaccine effectiveness in the under fours was only 49%, and less in older children but once again it was a less severe and a shorter illness in the immunised.

The second big study was from Professor Paul Grob of the Epidemic Observation Unit of the Royal College of General Practitioners and appeared on 13th June 1981 [5]. It studied in a similar way to Williams, with specialist nurses following cases of suspected whooping cough referred by interested GPs from the south-west Thames region. The patients were at the other end of the socio-economic spectrum from Williams's deprived population and the results showed a much lower incidence of hospital admission among the 658 cases studied. They were primarily investigating the benefit of antibiotics in whooping cough, which proved very difficult, but their observations allowed a detailed assessment of vaccine protection in home contacts. The vaccine protection in that challenging situation was 70%, and the reduced severity in the immunised was also once again confirmed.

All these studies were done differently, yet on balance confirmed that the vaccine reduced the numbers affected and the severity. Ditchburn's Shetland patients showed no benefit, but it was considered a very weak study.[3] The West Glamorgan study supported Professor Gordon Stewart's view, however, that social conditions were a significant factor in the epidemiology of pertussis.

The scales had definitely come down on the side of the vaccine working, but the amount of individual protection afforded was poor compared with what one hoped from a vaccine. What we were learning, that was to become a key issue for *Bordetella pertussis* prevention, was that there was a big and important difference between the amount of protection it gave an individual exposed to the infection, known as vaccine efficacy, and the amount of protection afforded to the whole population by having a high proportion of them immunised, which reduces spread very successfully. This latter phenomenon is called herd immunity colloquially, and technically it is called vaccine effectiveness.[4]

Meanwhile we were still waiting to know if the vaccine was safe.

References

1. Jenkinson D. Outbreak of whooping cough in general practice. BMJ. 1978,310:5/7–8. https://www.ncbi.nlm.nih.gov/pmc/articles/PMC1607029/.
2. Cherry JD. The epidemiology of pertussis and pertussis immunization in the United Kingdom and the United States: a comparative study. Curr Probl Pediatr. 1984 Feb 1;14(2):7–7.
3. Nicoll A, Gardner A. Whooping cough and unrecognised postperinatal mortality. Arch Dis Child. 1988 Jan 1;63(1):41–7.
4. Royal College of General Practitioners. Effect of a low pertussis vaccination uptake on a large community. BMJ. 1981;282:23–6.
5. Grob PR, Crowder MJ, Robbins JF. Effect of vaccination on severity and dissemination of whooping cough. BMJ. 1981;282:1925–8.

---

[3] We now know the effect of the vaccine is not long lasting. With the benefit of hindsight we can say his methodology was inappropriate.

[4] Because effectiveness has both an everyday meaning and a technical one, there may be unavoidable instances of ambiguity.

Chapter 7
1981. Whooping Cough Vaccine Is Very Rarely Harmful

The National Childhood Encephalopathy Study (*NCES*) published its findings in May 1981 [1]. This study group had been set up by the Joint Committee on Vaccination and Immunisation to investigate the frequency of brain damage after pertussis immunisation. It was a sub-committee under the chairmanship of Professor Alistair Dudgeon and the investigation was undertaken by Professor David Miller and his team in the Department of Community Medicine at the Middlesex Hospital Medical School. It commenced work in June 1976. Interest in the study was intense as anyone involved in pertussis immunisation was hoping for a positive outcome that would justify their belief it was safe and their continuing to encourage parents to accept it.

While it was ongoing many people seemed to be in the know that the study was having difficulty coming up with an answer, but this was good news as it was taken to mean that the numbers were not yet significant. The study was a case control study in which every time a child was admitted to hospital with brain inflammation (encephalitis) two randomly selected children from the same community, matched for age and sex, were identified for comparison of recent immunisation history. It was a comprehensive study involving all paediatric units in England, Wales and Scotland. Nothing like it had previously been attempted. By the time the results were published in May 1981 the answer in general terms was already known.

Before the *NCES* the best estimate of the incidence of brain damage after pertussis immunisation was based on what was known to doctors as the 'Yellow Card' system. Every doctor was supplied by the Committee on the Safety of Medicines (*CSM*) with a mailable self-sealing card, which happened to be bright yellow, on which to report adverse drug reactions. In practice it was usually just the more serious ones that were reported. Between 1964 and 1976 there had been 716 suspected vaccine reactions of various sorts reported by this means. In 1977 the *CSM* needed to provide a figure from these reports that estimated the likely frequency of brain damage, so that David Ennals, the Health Minister, could answer a parliamentary question about it. The number the minister gave out in Parliament was one for every

D. Jenkinson, *Outbreak in the Village*, Springer Biographies, https://doi.org/10.1007/978-3-030-45485-2_7

300,000 children vaccinated. This 'official' number was thereafter quoted by doctors who needed to quantify the risk for concerned parents and others, but it was common knowledge that the yellow card system was probably very inaccurate. That was why the *NCES* was required. The scary 1974 report from Great Ormond Street Hospital that frightened everyone was unable to quantify the risk but implied it could be much higher than the official figure.

The *NCES* reported in 1981 after analysing 1000 cases of encephalitis. It confirmed that reactions severe enough to cause brain injury were more frequent after diphtheria, tetanus and pertussis immunisation (*DTP*) than after diphtheria and tetanus immunisation (*DT*) alone. There were 35 permanently damaged children who had had pertussis vaccine in the previous seven days. The *relative risk* was 2.4. The encephalitis rate was 1 in every 110,000 injections, but because the majority fully recovered, the risk of permanent brain injury was 1 in 310,000 injections. Because each child requires three injections the risk could be said to be as high as one in every 100,000 children immunised, but as a reaction would normally cease the *DTP*[1] course the risk probably lay between these two very small numbers.

All the study could confirm was an association with *DTP* but no causal relationship. There was nothing to differentiate possible damage associated with pertussis vaccine from damage from other causes.

Looking at the risk a different way, it meant that a single-handed GP immunising all his new born patients would see a case of brain damage associated with *DTP* once every 2000 years! Without pertussis immunisation that same GP would witness several young infants dying of pertussis in a single practice lifetime of 35 years. The bottom line was that pertussis vaccine definitely saved many lives.

The *NCES* was criticised and reanalysed by many people after it was published but nobody found any fundamental flaws or errors that changed the general conclusion. There was a possibility that the risk might be greater if contraindications were not observed properly, but that should not apply where there is a good medical service. The *NCES* continued its follow-up of the children and half the controls where it was possible (about 80%) for 10 years and reported the outcome again in 1993 [2]. The *relative risk* was revised to 5.5 but it was pointed out that because the actual number of children considered affected was so small, that statistically the relative risk could be several times higher or lower, but still the bottom line was the same. The vaccine saved lives and the risk was very, very small.

It is now known that whole cell pertussis vaccine spotlights individuals destined to develop infant epilepsy and its complications. Any fever-associated challenge would have eventually brought the same result, and there are always many in early childhood. The association was real but only because events that were going to happen anyway were brought forward in time. Whole cell pertussis vaccine does *not* cause brain damage.

It is disappointing to look back at that time and see how persistently sections of the media pursued the anti-vaccine story. It started with a certain amount of reason

---

[1] Although the same cessation had possibly been applied to those included in the study.

in 1974 but when the consequences emerged and fears were shown to be unfounded, even after ten years, a sensational headline could still sell newspapers (Fig. 7.1).

It took many years for *DTP* to be re-accepted as the vaccine to give all infants and only passed the 90% level in 1992 from its lowest of 31% in 1976. It finally caught up with diphtheria and tetanus at 94% in 2005, even though other measures were being taken to reduce the risk of reactions. One of these new measures was applied in 1990 with the introduction of an accelerated immunisation schedule, given at two, three and four months, rather than three, five and eleven months, in order to give full protection at as young an age as possible in this most susceptible age group. Another advantage of the early completion age was that younger children are less susceptible to brain injury, so reactions were less severe. The other measure taken was a change to a purer (acellular) pertussis vaccine that caused fewer reactions, but not until 2004 in the UK.

Back in 1981 when the uptake was about 45% nationally, we had a particular problem in Keyworth because our uptake rate was quite a lot lower. Health visitors Gwen and Jackie were the people whose responsibility it was to get the consent forms signed for immunisation, and it invariably included a discussion about whether to have pertussis vaccine or not. Most of their clients were educated and informed enough to have been scared off having it by the media publicity which was very persuasive. The health visitors could only explain that reactions were rare and that whooping cough was a terrible disease for a baby to get. Parents would then often point out that by the time their baby's immunisation was effective, at about a

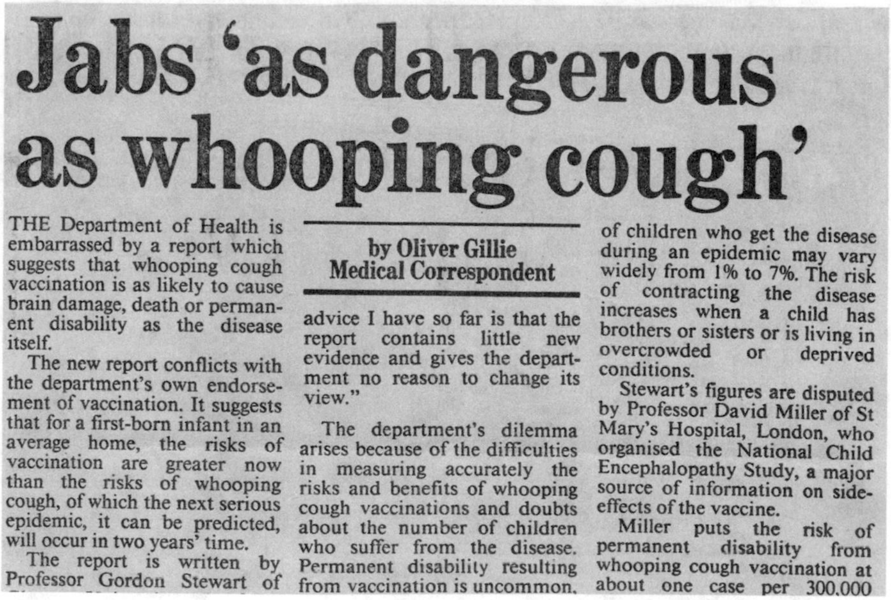

Jabs 'as dangerous as whooping cough'

THE Department of Health is embarrassed by a report which suggests that whooping cough vaccination is as likely to cause brain damage, death or permanent disability as the disease itself.

The new report conflicts with the department's own endorsement of vaccination. It suggests that for a first-born infant in an average home, the risks of vaccination are greater now than the risks of whooping cough, of which the next serious epidemic, it can be predicted, will occur in two years' time.

The report is written by Professor Gordon Stewart of

by Oliver Gillie
Medical Correspondent

advice I have so far is that the report contains little new evidence and gives the department no reason to change its view."

The department's dilemma arises because of the difficulties in measuring accurately the risks and benefits of whooping cough vaccinations and doubts about the number of children who suffer from the disease. Permanent disability resulting from vaccination is uncommon.

of children who get the disease during an epidemic may vary widely from 1% to 7%. The risk of contracting the disease increases when a child has brothers or sisters or is living in overcrowded or deprived conditions.

Stewart's figures are disputed by Professor David Miller of St Mary's Hospital, London, who organised the National Child Encephalopathy Study, a major source of information on side-effects of the vaccine.

Miller puts the risk of permanent disability from whooping cough vaccination at about one case per 300,000

Fig. 7.1 *Sunday Times* 17th June 1984. The article itself was more balanced but most people would just read the headline

year old, it would be past the danger period, which was true. The health visitors would then point out that babies usually caught it from older siblings, so it was actually the next baby that was going to benefit. They would also go on to explain about how herd immunity stopped the disease spreading, which was possibly a much greater benefit. It inevitably became rather complicated, and in the end, parents often just followed their instinct to play safe by declining pertussis vaccine or, having weighed up whether to trust their health advisor or their newspaper, decided to place it in the latter. After all, none of them knew anyone who had lost a child to whooping cough.

As a GP, I too often had this conversation. Realising I could not expect them to simply follow my advice on trust, I sometimes made altruism the issue. I would say, trying to hide the profound irony, that as parents trying to do the very best for their child, they should not have their child immunised but persuade all other parents to have their children immunised. I think this sometimes got the point home.

After the big 1977–9 outbreak I think people started to understand the consequences of the fall in the immunisation rate and more were prepared to accept the argument for immunisation, but the health visitors and I wanted to improve the rate as fast as possible and we tried to think of ways we could do it. I cannot remember whose idea it was but we analysed the uptake figures in our patients and found that families tended to stick to the decision that they made for their first child for all their children, thinking, I suppose, that whatever was good for the first would probably be good for the next, and so on. Logically then, efforts to change behaviour should be best focussed on the parents of firstborn children.

The graph in Fig. 7.2 shows that for firstborns (blue) born in 1980, their uptake of pertussis vaccine was 92% compared with 73% in second and subsequent children (brown). Both are very much higher than the 41% for England and Wales (green) for all birthranks.

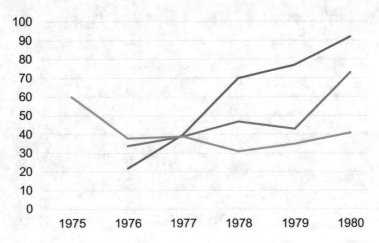

Fig. 7.2 Pertussis vaccine uptake as percentage in Keyworth firstborn (blue), others (brown), and England and Wales (green) 1975–80

This seemed a useful observation that could be translated into policy, so we accordingly wrote a letter with our findings to the *BMJ*, which they printed on 18th April 1981 (Fig. 7.3).

There was a paper in *The Lancet* in early 1982 [3] that appeared when whooping cough cases were rising rapidly again. The rise was expected because epidemics of whooping cough occur roughly every four years and this had been true as far back as anyone could remember without a convincing explanation. Paul Fine and Jacqueline Clarkson worked at the London School of Hygiene and Tropical Medicine and were familiar with the mathematics of epidemics. They had looked at whooping cough notifications and thought that the four yearly cycles that were persisting in spite of the big drop in vaccination were telling us something important about the infection that was hitherto unrecognised.

The cycling pattern of some infectious diseases is well known. Measles epidemics occurred every two years until immunisation was introduced in 1967. I am just old enough to remember an outbreak of measles when I was a locum for doctors Marcus and Kapur in Swansea in 1968. (As a poor hospital doctor I had to work in my summer holiday to make ends meet.) I wrote out a lot of measles notification forms which helped maintain their practice income.

These patterns occur because once a person has the infection, they spread it to others around them who are susceptible and they in turn catch it and pass it on until there are insufficient susceptibles to maintain the spread and the epidemic then fades away. When the number of susceptibles builds up again, with new children in the case of measles, to a level where one person can spread it to more than one other, the scene is set for another outbreak, and so the cycle repeats. Fine and Clarkson pointed out that following adverse publicity, the acceptance rate for pertussis vaccine had dropped to about half what it was before, and therefore should have doubled the number of susceptibles, which should have halved the inter-epidemic period to two years rather than four but it hadn't. The probable explanation,

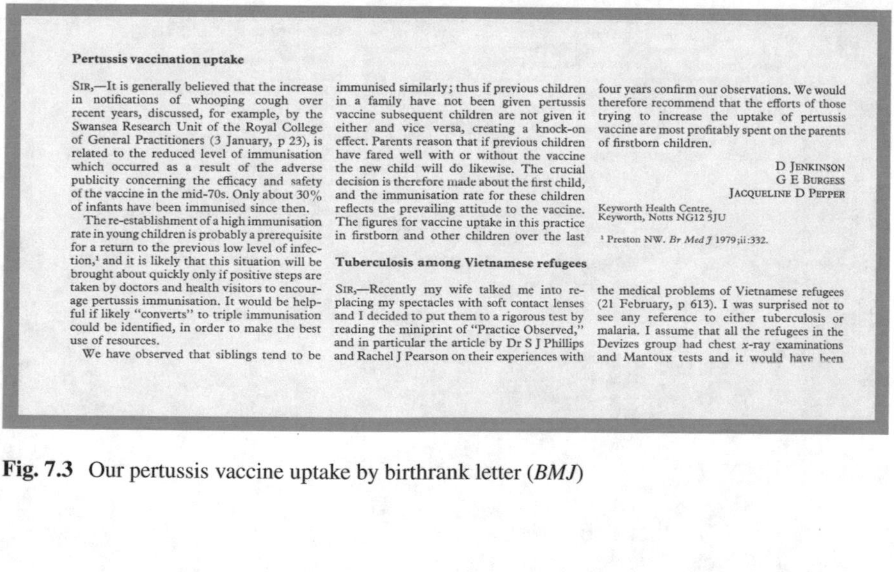

Pertussis vaccination uptake

Sir,—It is generally believed that the increase in notifications of whooping cough over recent years, discussed, for example, by the Swansea Research Unit of the Royal College of General Practitioners (3 January, p 23), is related to the reduced level of immunisation which occurred as a result of the adverse publicity concerning the efficacy and safety of the vaccine in the mid-70s. Only about 30% of infants have been immunised since then.

The re-establishment of a high immunisation rate in young children is probably a prerequisite for a return to the previous low level of infection,[1] and it is likely that this situation will be brought about quickly only if positive steps are taken by doctors and health visitors to encourage pertussis immunisation. It would be helpful if likely "converts" to triple immunisation could be identified, in order to make the best use of resources.

We have observed that siblings tend to be immunised similarly; thus if previous children in a family have not been given pertussis vaccine subsequent children are not given it either and vice versa, creating a knock-on effect. Parents reason that if previous children have fared well with or without the vaccine the new child will do likewise. The crucial decision is therefore made about the first child, and the immunisation rate for these children reflects the prevailing attitude to the vaccine. The figures for vaccine uptake in this practice in firstborn and other children over the last four years confirm our observations. We would therefore recommend that the efforts of those trying to increase the uptake of pertussis vaccine are most profitably spent on the parents of firstborn children.

D JENKINSON
G E BURGESS
JACQUELINE D PEPPER

Keyworth Health Centre,
Keyworth, Notts NG12 5JU

[1] Preston NW. *Br Med J* 1979;ii:332.

Tuberculosis among Vietnamese refugees

Sir,—Recently my wife talked me into replacing my spectacles with soft contact lenses and I decided to put them to a rigorous test by reading the miniprint of "Practice Observed," and in particular the article by Dr S J Phillips and Rachel J Pearson on their experiences with the medical problems of Vietnamese refugees (21 February, p 613). I was surprised not to see any reference to either tuberculosis or malaria. I assume that all the refugees in the Devizes group had chest x-ray examinations and Mantoux tests and it would have been

Fig. 7.3 Our pertussis vaccine uptake by birthrank letter (*BMJ*)

according to them, was that the vaccine didn't stop children getting infected, but it did stop them getting the disease associated with the infection.

An analogy to illustrate the theory might go like this. Think of a game of cricket. In that game, hard balls fly around in ways that would sometimes cause broken legs and jaws if batsmen did not protect themselves with pads and helmets. The vaccine works like the pads and helmets; it stops the harmful agent doing damage but does nothing to reduce the number of dangerous balls flying around. A good vaccine should stop the balls flying around too.

If Fine and Clarkson's view was correct, this would have important implications for vaccine development and use. *Bordetella pertussis* only affects human beings, so in theory, if the organism can be eliminated, it would be gone for ever, just like smallpox, and has almost been achieved with polio. But if the vaccine we use only prevents the disease and not the infection, then there could be no hope of eradication by its use. A completely different vaccine would have to be developed for eradication. It might seem to be a straightforward issue, but now, nearly 40 years after this paper was published, we know that their theory does not accurately describe what really happens, but they were on the right track. Nevertheless, how the vaccine works, and how *Bordetella pertussis* spreads and causes the symptoms it does, is a bit clearer now but much remains to be discovered.

References

1. National Childhood Encephalopathy Study, Miller DL, Ross EM, Alderslade R, Bellman MH, Rawson NS. Pertussis immunisation and serious acute neurological illness in children. BMJ. 1981;282:1595.
2. Miller D, Madge N, Diamond J, Wadsworth J, Ross E. Pertussis immunisation and serious acute neurological illnesses in children. BMJ. 1993;307:1171.
3. Fine PEM, Clarkson JA. The recurrence of whooping cough: possible implications for assessment of vaccine efficacy. Lancet. 1982;319:666–8.

Chapter 8
1981–83. The Second Outbreak. How Many Cases Are There Really?

My work as a GP changed radically in the almost 38 years I spent in that role. It was usually demanding and busy, although I remember the school summer holidays could sometimes be a slack time in the early 1980s. Many people were away on holiday and there were some days when there were no home visits to do, which was unusual but wonderful. It all gradually changed as the population lived longer and standards and expectations became higher. We then worked nearly 24/7 but under less pressure. Many GPs still work long hours and all are under greater and constant pressure, but for the most part they now have regular and reasonable time off.

When I think back to how things were medically when I qualified in 1967, I feel a certain amount of horror. People now have little idea how poor medical care used to be compared with today, especially for the elderly, who had little hospital care. Hospital beds were scarce and guarded jealously by the junior doctors who controlled them and who dared not use them for the sick elderly as they would incur the wrath of their consultant. There was no such thing as a manager then and patients would stay in hospital until the consultant was happy for them to go home, so the wards were usually full. Many hospitals had introduced 'geriatric' wards, often bed-blocked with patients with dementia or general frailty, and once a hospital had such wards the medical consultants would forbid their junior doctors on the ward to admit any patient over 65. Yes 65! They were the business of the geriatricians! The only way into a hospital bed at that time was via a telephone call from the GP to the junior hospital doctor 'on take' who would often say, if it were a medical (as opposed to surgical) problem, 'No, you must ask the geriatric team'. Over 65 and you knew you were probably wasting your time trying because they never had any beds. Sometimes in desperation you would have to abandon the patient and tell them to ring 999 and say it was an emergency. The ambulance men (there were no paramedics then) would have to take the patient to casualty where the patient would be seen to be seriously ill and urgently admitted by the same doctor who refused on the phone! What a game! For a GP in that era, getting an 'elderly' patient admitted was one of the most stressful things we had to do.

© The Editor(s) (if applicable) and The Author(s), under exclusive license to
Springer Nature Switzerland AG 2020
D. Jenkinson, *Outbreak in the Village*, Springer Biographies,
https://doi.org/10.1007/978-3-030-45485-2_8

In late 1970, while I was waiting for a boat to take me to Africa, I did a GP locum job in Winsford in Cheshire. Winsford was once a small village but now had a new large housing estate which had been built to accommodate the overflow and rehousing of Liverpool's inner city. The surgery was in an end house in a terrace of new houses on this estate. There were a lot of pregnancies and the ones where complications might be expected would be booked for hospital delivery. A previous caesarean section automatically came into this category for many reasons, one being the possibility of a repeat section being needed, and also there was the risk of the uterine scar rupturing during labour, something that was usually fatal for the baby and potentially so for the mother. My letter of referral for such a patient was returned by the obstetric consultant in Chester, explaining that he had insufficient beds to book her for hospital delivery. We were presumably supposed to conduct this dangerous delivery in her home!

We should be grateful that the NHS has improved so much since then. The introduction of managers overcame some of the issues like the ones I have mentioned. It is not just down to money but mainly to attitudes. Patients' needs are now the centre point of medical care and supported by politicians. Doctors and other health workers are becoming small cogs in the big healthcare engine.

The ability of an individual doctor in the NHS to undertake research has been affected by these same changes. Plenty of research is being done but it has to be planned, organised and financed, and it usually takes a large team. I sometimes wonder if it is possible for a modern GP to research in the way I have done. I hope so.

*****

I had had plenty of time after the previous outbreak to reflect on what I had done and what I was going to do in the future about it. I thought myself a bit of an expert on the clinical diagnosis of whooping cough, and therefore I should continue to diagnose all the cases I could. Nobody knew what the long-term future would be for the disease and immunisation practice and I wanted to be part of the action whatever it might be. It seemed a good disease to be involved with because it came in waves which allowed one to have a 'breather' before the next wave. David Hull had casually advised, 'Keep recording' in 1978, and knowing he was a most insightful person, I took it seriously, and I am exceedingly glad that I did.

There had been changes in the Keyworth practice. The old guard had been replaced by a new guard of formally trained GPs. Manson Russell, the doctor who set up in Keyworth soon after the start of the NHS in 1948, had retired and gone to live in Ireland, and Rowan Stevenson, a great nephew of the author Robert Louis Stevenson, went soon afterwards. Andy Watson, who was a product of Nottingham's new medical school, spent a few years with us before leaving to specialise in orthopaedic medicine, but Clive Ledger took his place, and soon afterwards Andrew Wood joined us. Clive and Andrew both came via the excellent Airedale Training

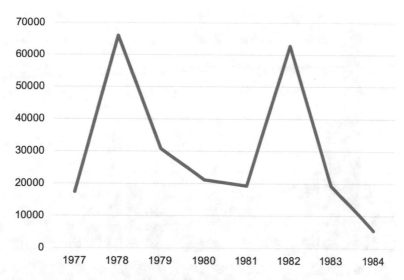

Fig. 8.1 Notifications of whooping cough in England and Wales 1977–1984

scheme in Yorkshire and we worked well as a team right from the start. Five years later in 1988 Andrew became a GP trainer himself, and for a while after, the practice benefitted from extra pairs of hands in the form of trainee GPs, later called registrars. They all quickly became educated about whooping cough.

*****

As expected, right on time for the four-yearly cycle, the first new cases came along in August 1981, and took off when the school term got under way.

The size and shape of the epidemic in 1981 to 1983 was very similar to the one of 1977 to 1979 for England and Wales as a whole (Fig. 8.1). Smaller patterns though are hidden within the large one. The pattern in Keyworth was rather different in the two outbreaks. In the first one it was like an explosion that reached a crescendo in the November of 1977 then quickly faded away. In the second (1981–83) there were two distinct peaks a year apart and not nearly as large individually as in the previous outbreak in 1977–9 (Fig. 8.2).

I think the double peak pattern can be explained in terms of the number of susceptibles. The total number of cases in 1981–3 was a third of the 1977–9 total. This implies there were less susceptibles in 1981–83. That makes it more difficult for a single case to pass it on to more than one person and start a chain reaction. Because the number of susceptibles within range of an individual sufferer will vary by chance, many small clusters of cases will quickly fizzle out. Both the increases seen in the above chart start at the end of the school summer holiday and reach a peak in the September, at the start of the school term. It was just the same at the start in 1977. Not what you would expect if schoolmates mixing in class were the source of susceptibles. If that were so, you would expect cases to begin appearing after the

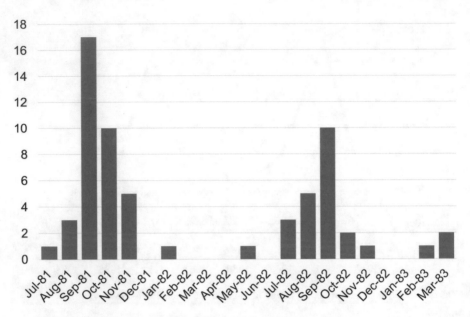

Fig. 8.2 Keyworth whooping cough cases by month 1981–1983

start of the term, not before. The starts of the 1981 and 1982 little outbreaks were well before the start of school. Why this should be is just another question to add to the long list of questions about whooping cough that keeps me fascinated.

I found seven adults that caught whooping cough this time, out of over 60 cases, and one five-year-old developed pneumonia. Six-year-olds had the highest incidence whereas four years previously it was three-year-olds, suggesting that they somehow managed to avoid it at that time, perhaps because they were too young to go to school or playgroup.[1]

The first column in Fig. 8.2 represents a boy 5ym[258] whom I had prevented from getting whooping cough in 1978. He started coughing at the end of July 1981 and coughed with typical whooping cough for six weeks. He had not been immunised and he did not have a swab taken. Back in February 1978 when he was two years old, his older brother 4ym[191] caught whooping cough and was confirmed with a swab taken on February 23, 1978. 5ym[258] himself had been started on erythromycin two days previously on February 21, 1978 as a preventative and he never developed whooping cough at that time. Although this case is an example of possible prevention it is not definite because when he caught it at five, it was not proven with a positive swab. I have had several instances of similar scenarios but always fallen short of absolute proof. Nevertheless, it is normal practice to give an antibiotic like this to susceptible contacts of whooping cough but always difficult to know how often it is actually prevented. In this case I probably prevented him

---

[1] With the benefit of hindsight we would now say it was the immunisation wearing off, but we were not supposed to think that then!

getting it at two, which may have allowed him to get it at five and carry it into his school. On the other hand, whooping cough at two is much more unpleasant and difficult.

✳✳✳✳✳

It was thought that there were two kinds of GP, notifiers and non-notifiers of infectious diseases. The Medical Officers of Health who counted the numbers knew that the same GP names kept appearing on the forms (Fig. 8.3), but a lot of names never appeared at all. It was accepted that although certain diseases were supposed to be notified, there was a wide variation between doctors and diseases. It would be very difficult to change GP behaviour, and if it succeeded it would render previous statistics difficult to interpret. It was thought pragmatic to leave things alone and estimate the real figure, taking into account the likely attitude of GPs to the notification of any particular disease, or use another means of counting altogether such as laboratory data on the frequency of infections.

My experience in the previous epidemic of 1977–9 had told me that I had seen far more cases than were being notified in Nottingham as a whole and therefore I wanted to find out if cases were not being diagnosed or not being notified. If it were the former, some educating needed to be done. If the latter, then in a sense, all was well.

On top of the notification habit there was the diagnostic ability. Any doctor who had qualified since about 1960 was unlikely to be familiar with the disease, but there

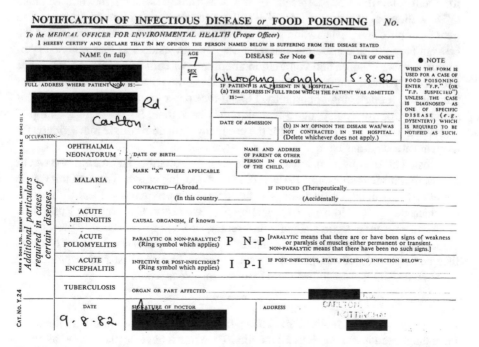

Fig. 8.3 The standard form used to notify an infectious disease or food poisoning in 1982

were plenty of doctors in Nottingham who had qualified before then, including many from the Indian subcontinent, who travelled to the UK to work in the NHS and who would definitely have been familiar with whooping cough.

I wanted to find out what proportion of diagnosed cases was notified. I decided to survey all Nottingham doctors to find the answer, so in mid-1982 I started planning a postal questionnaire. Previous epidemics suggested that the peaks come about September time, so I set September 1982 as the month of the survey. The questionnaire itself was very simple. It took the form of a letter addressed personally to every GP in the Nottingham Health Authority area and asked, 'How many new cases of whooping cough did you diagnose in September 1982?'. The number had to be written on the letter and returned in the pre-addressed and stamped envelope. The letter also requested that only patients living in the Nottingham area be included, as some practices covered adjacent areas too, and I emphasised that their normal notification practice should not be changed.

The Family Practitioner Committee (*FPC*) was the body that oversaw GPs at that time, and I obtained the names and addresses of all the listed GPs from the Nottinghamshire *FPC*. Doctors were not allowed to work as GPs in the NHS unless they were included in the *FPC* list. The *FPC* also kept a list of all the patients registered with each of these doctors. It is much the same system today.

So a rainy Sunday afternoon was spent addressing 292 envelopes and sticking stamps on them. The cost of the stamps was met by Abbott Laboratories Ltd., a well-known pharmaceutical company. I had made a video about whooping cough for them after the first outbreak. In it I was interviewed by the late Michael O'Donnell, a popular doctor-journalist and broadcaster, about how to diagnose and treat whooping cough. The main point at that time being that erythromycin was the best antibiotic to stop sufferers passing it on. The downside was that generic erythromycin caused nausea and vomiting in about ten per cent of people. Abbott made a form of erythromycin that was better tolerated and was the preferred choice, so for them it was a marketing opportunity. I had a nice day out, met a celebrity and was given £400 for my trouble, quite a lot of money back then. So 384 stamps on top of that was a minuscule investment for the company. I made a couple more videos for them at various times later.

I posted all the questionnaires off at the end of September and within two weeks 244 were returned reporting 518 diagnoses of whooping cough. The next job was to count up the official notifications. The notifications were made to local authorities, of which there were four in Nottingham City, Broxtowe, Gedling and Rushcliffe. Each one kept their records slightly differently, so I had to visit each to present my credentials in order to get the information. I ended up with a combination of typed lists and photocopied notification forms covering both August and September. I waited until October to collect them to allow for late arrivals.

*****

A few years later I realised that I had personally neglected to notify all the cases I had diagnosed in the previous 12 months. There must have been about 30 or more. I was worried that I might lose credibility for my research of the disease if I had not notified cases. Ignoring the fact that the system was for rapid alerting, I filled out the forms and sent them off. They ended up on the desk of microbiologist Dr. Richard Slack, who must have been livid to receive them as not only were they pointless but would mess up his statistics badly. Being the charmingly polite man he was, he sent me a bitingly sarcastic letter that I richly deserved for bucking the system. I am sure I remained an irritation to Richard, as the number I notified was way beyond normal and must have distorted the Nottingham figures for years on end.

I returned from a trip to the Amazon in 2001 with a nasty gut infection called *campylobacter.* Richard phoned me to confirm it and to tell me I was the first person he had known to notify his own disease! I blame it on the tapir that ran loose in the kitchen.

*****

I needed to make a small adjustment to the notification count because there are two dates on each notification form. The principal one is the date the disease started, and the other is the date of notification. Neither of these dates was the date of diagnosis, which was what I had asked for on the questionnaire. It can take two weeks for the cough to sound like whooping cough. Then there is a possible delay before it gets notified. The average difference between whooping cough starting and being notified was ten days, so I used the midpoint of five days as the adjustment, although it made very little difference to the actual result. Allowance was also made for the few doctors who made no return, by assuming they saw the average number seen by the responders. That worked out at 620 cases diagnosed by the GPs. They notified 116 of them which works out at 18.7%, roughly one fifth.

No previous researchers had used a method like this but different estimates over the years had come out at about 20%, very much the same. A big question mark had been recently raised by experts who thought GPs might be more likely to notify in an epidemic when there is a lot of publicity about it. This view was supported by evidence from certain officially designated recording practices reporting all infectious diseases to the Royal College of GPs statistical centre, separately from the notification system. In the previous epidemic these figures had shown a smaller rise than notifications, suggesting a larger proportion are notified in an epidemic. My result suggested that the proportion notified remained the same in an epidemic. I was more confident about this when it was confirmed that the weeks covered in September 1982 included the maximum weekly rates of notifications in 25 years and were not exceeded thereafter.

The 244 doctors who replied gave widely different numbers diagnosed, which illustrated my belief that some diagnosed it better or more readily than others. 109 reported none at all, ten reported eight cases, one reported 12, and one 24 (not me! I was a ten). The remainder were somewhere in between.

The official number of notifications of whooping cough in 1982 in England and Wales was 65,815. If you apply a correcting factor based on my findings it works out at over 350,000 cases. The number I was personally diagnosing in this outbreak was five times the national rate. It follows that I was therefore diagnosing at the same rate as other Nottingham GPs but they were only notifying 20% of them. A

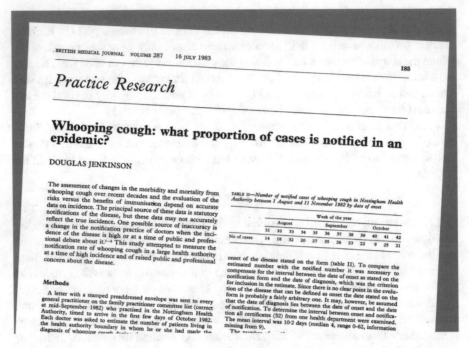

Fig. 8.4   The published notification rate paper in the *BMJ*

decade or so later, most of these GPs had retired and with them went the ability to diagnose whooping cough.

The next job was to write it up, and ever optimistic, I sent it to the *BMJ* again. They accepted it in June 1983, and it was published on 16th July [1] (Fig. 8.4).

I didn't expect any feedback or comment on a very uncontroversial paper, so I was not surprised when there was very little.

In Keyworth it had all gone quiet whooping cough wise by the end of 1982. There were three cases in 1983 and none in 1984. It all started again in 1985.

Reference

1. Jenkinson D. Whooping cough: what proportion of cases is notified in an epidemic? BMJ. 1983;287:185–6.

Chapter 9
1985–87. The Third Outbreak. The Search for Subclinical Infection

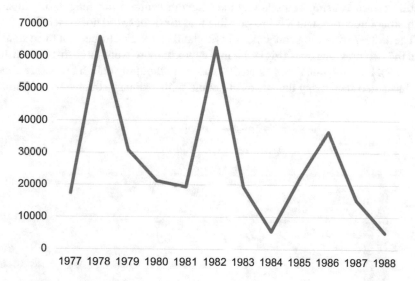

Fig. 9.1 Whooping cough notifications by year 1977–88 in England and Wales

Diseases are named for different reasons. Those that have been known for a long time are most likely descriptive of the symptoms (what the patient feels) and signs (what others see) that go with them.[1] They were named long before their cause was known and most we now understand are caused by micro-organisms. Because different micro-organisms invade our bodies in different ways, the symptoms associated with each tend to be distinctive. We all know what the common cold is and can

---

[1] Although it can be important to separate 'symptoms' from 'signs', in everyday speech the word 'symptoms' is taken to include both.

© The Editor(s) (if applicable) and The Author(s), under exclusive license to
Springer Nature Switzerland AG 2020
D. Jenkinson, *Outbreak in the Village*, Springer Biographies,
https://doi.org/10.1007/978-3-030-45485-2_9

diagnose it in ourselves easily. We also know that a cold can be quite severe or very mild, and it is probably true to say that all infectious diseases vary in the severity of the symptoms they cause. If somebody has a bad cold you probably want to keep out of their way because they are sneezing a lot and you believe that is the way it spreads, and you would be right.

Whooping cough was obviously named because of the whooping noise that often went with it but not everybody makes that noise. In the past it was also called chin-cough, a term probably derived from the old Scottish name for it which was 'kink-cough' or 'kinch-cough', which means a gasping cough. That is a far better name because gasping for air is the consequence of a paroxysm which is itself the characteristic that distinguishes it from other coughs. It was seen to pass from person to person, usually in childhood. Some descriptions from long before immunisation suggest that many people seemed to escape whooping cough altogether. It was only first described causing outbreaks in the fifteenth century, not long ago compared with other infections, and it is just possible that the disease did not exist before that.[2]

The 1985-7 whooping cough peak in England and Wales turned out to be smaller than the two previous ones (Fig. 9.1). In Keyworth there were three distinct clusters (Fig. 9.2). By 1984 I had seen about 300 cases of whooping cough I was sure about, but I had also seen many thousands of people with ordinary coughs. By and large I

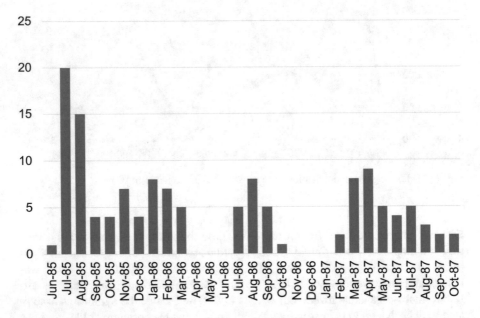

Fig. 9.2 Keyworth whooping cough cases by month 1985–87 showing three main clusterings

---

[2] The animal respiratory infection 'kennel cough' is caused by *Bordetella bronchiseptica* which can infect many animal species but not humans. It has more DNA than *Bordetella pertussis*, which is probably a diminished version only able to infect humans.

did not have a problem telling the difference but often only after several weeks had passed and using the benefit of hindsight. Patients with any paroxysmal cough had to be suspected of having whooping cough and the vast majority kept up that kind of cough for three weeks or more, removing most doubt. There had been a few who had a paroxysmal cough that only lasted one to two weeks and a small proportion of these had swabs taken. One at least had come back positive. I also knew that many household contacts of whooping cough had ordinary coughs. A few of these had had swabs taken too, but none had been positive.

It concerned me to think that there could be many more cases than I was diagnosing if it frequently caused mild symptoms that made it impossible to distinguish it from other causes of cough, such as common wintertime viruses. It would mean that efforts to control it by isolation or antibiotics would be largely futile and would open up the whole vaccine effectiveness question again. This whole question of subclinical or asymptomatic infection needed serious consideration, especially in view of Fine and Clarkson's theory in 1982 that the vaccine did not prevent infection, just the disease.

I confess I felt sceptical about subclinical infection being an important issue, largely because I had watched the start and spread of two outbreaks in Keyworth, and on both occasions the route of spread could be explained by following cases of clinical whooping cough; that is, patients with a minimum of three weeks of paroxysmal coughing. My observations were telling me it was they, the ones with clinical whooping cough that were the transmitters. But we know you *can* have whooping cough with less than three weeks of it, so there must have been some with a shorter duration of paroxysmal coughing, and perhaps even ordinary coughing.

An ordinary cough with no paroxysmal coughing and positive swab for *Bordetella pertussis* is called *subclinical* infection, because it is the paroxysmal cough that is the characteristic of the disease. A three-week duration definition maximises the probability of being correct. If there is a positive swab with no symptoms at all, it is called *asymptomatic*. I planned to look for subclinical infection in the next outbreak, wondering if it would change my understanding of whooping cough by showing there was a substantial number of them. I was postulating four classes of *Bordetella pertussis* infection:

1. Three weeks or more of paroxysmal coughing.
2. Less than three weeks of paroxysmal coughing.
3. Ordinary coughing.
4. No symptoms.

I was going to go looking for the first three but not number four.

I could not possibly have done it on my own. My work schedule was far too tight in a busy general practice and I had the additional load of an attached medical student most of the time since I had taken on a part-time lectureship in general practice. All the other 11 part-time lecturers had negotiated reduced hours in their practice to compensate for the extra work, but I had just tacked it on as we were just too busy to allow any slack. Fortunately, health visitor Jackie Pepper had the same curiosity and drive to investigate whooping cough as myself and she volunteered to do the donkey work of case-finding that we foresaw. Jackie was by this time well known in the village by parents and schools because of her job and was highly respected. That would make her task considerably easier when she had to ring parents to find out if their kids had a cough or find out about seating positions in schools etc. to trace possible contacts.

We became aware of the first case 8ym[348] in July 1985. He probably caught it when he was staying away from home in Hinckley at the end of May and beginning of June. His symptoms started on June 15th. He has the letter 'A' at the apex of the top cluster in Fig. 9.3.

All three outbreaks in Keyworth had been started by an identifiable source after spending time outside the area. I suppose this is not surprising but why the summertime in June, July and August?

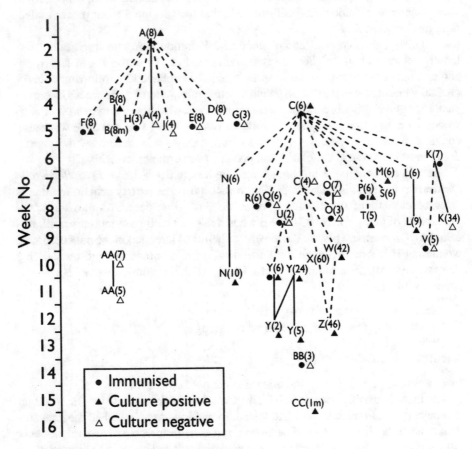

Fig. 9.3 The perceived pattern of spread of clinical whooping cough 1985 in Keyworth. Each family has a separate letter and the age is in brackets. Solid lines connect patients in the same family. Broken lines are non-family contacts. Other symbols are explained in the key. Numbers on the left are weeks since the start in June (*BMJ*)

Within four weeks four of his classmates and his brother 4ym[347] and several other contacts had started with it but that cluster seemed to go no further.

The second cluster starts in family 'C'. There were three children in family 'C'. Their parents ran a pub and the children mixed freely with the staff, most of whom also caught whooping cough. The middle child, a six-year-old girl, appeared to be responsible for most cases including her four-year-old brother who didn't do so badly at spreading it either! The only connection we could find between 'A' and 'C' was that in family 'C' there was also an eight-year-old immunised boy who was in the same class as 'A', but he had apparently had no cough! If he was responsible, he would have to have been an asymptomatic transmitter. It is possible that that is what happened, but if so, the bacteria would presumably have to have been transmitted by some other means than a cough. By transfer of saliva or nasal mucus probably. Easily accomplished when children are intermingling. On the other hand, perhaps he had a mild cough of short duration that did not register with any adult as signifying anything. It has to be borne in mind that we were dependent what we were told and had no way of verifying most of it. No doubt there were many things that were forgotten or ignored or distorted that might have changed our perceptions. There are many factors that interplay in the spread of infection and we do not fully understand the transmission of whooping cough.

Jackie and I had been diligently taking swabs from anyone with a cough that had been in contact with whooping cough. There had been a lot of publicity again in the media about the expected outbreak and now it was well known there were cases in the village. Some children were brought along because they had a cough and had been in contact, and their parents were keen to find out. Many kept away because they did not want to find out, but Jackie managed to discover a lot of them and do the necessary persuasion. We kept it up for three months.

Ultimately, we did not get a single positive swab from children with ordinary coughs, or from six contacts with no symptoms at all. We took 102 swabs from those with a cough and 17 of them were positive. All 17 had at least three weeks of paroxysmal coughing. There were another 16 with whooping cough who had negative swabs and a further six with whooping cough who had no swab taken. In this tiny study we had failed to find any evidence to support the hypothesis that subclinical infection occurred. But from other evidence we knew it did happen occasionally, but possibly not enough to change our overall understanding of it. I could still rely on clinical diagnosis to find most of the cases.

Parents, however, seemed to have a quite different perception of the disease. Most of them were surprised to have their child diagnosed with whooping cough because they had been led to believe from the media publicity that it was a serious illness, and it was still perceived as that by most of the medical profession too. Efforts were still being made to raise the immunisation rate, which was still only 65% in England and Wales in 1985, and it was being done by emphasising its seriousness in babies. It was not pointed out that it could go unrecognised in older children if they were perhaps only getting four or five paroxysms a day, largely because this aspect was not widely appreciated then, and still isn't. Most parents when given a diagnosis of whooping cough said the same thing. 'I thought it was a serious disease'.

During this outbreak we had made a particular point of asking about a catarrhal phase in the early stages of the disease. All the descriptions of whooping cough I had ever read, some going back 200 years, and most of those I still read today say whooping cough starts like a common cold with a runny nose they call the catarrhal phase, sometimes with sneezing too and voice changes. They typically say, 'Over the next week to 10 days a dry cough develops becoming gradually paroxysmal'.

Right from the early days it had struck me that such a catarrhal phase was frequently not mentioned by parents but I assumed that was because it was forgotten or not considered important, but then when I started seeing adults with it and I was able to question them in detail, I was getting stories of a really bad sore throat rather than catarrh. In September 1985 medical student Debbie Davey, who was on a month's attachment to the practice, questioned 25 parents in detail about their child's symptoms. Only nine thought there had been a catarrhal phase.

The most obvious explanation was that the symptoms simply varied from person to person, but whooping cough is an enigmatic disease with many mysterious aspects to it and I felt, and still do, that there are clues to the answers to some of these enigmas in the observations we make, if only we can work them out. I still have no firm evidence, but the frequent apparent absence of a catarrhal phase supported the idea in the back of my mind that it was sometimes a pre-existing viral cold that caused the catarrh, and the inflammation underlying it might have predisposed to acquiring *Bordetella pertussis* infection.

Bordetella pertussis bacteria invade the cilial layer that lines the surface of the epithelial cells[3] of the nasopharynx, trachea and bronchi. They cause coughing and sometimes sneezing and sore throat in a typical attack. The infection does not cause a productive cough like other respiratory infections unless there is secondary bacterial infection. So why should it produce nasal catarrh?

Another mystery for me is why it seems to cause intense little outbreaks that then fizzle out, just like all the cases started by 'A' in Fig. 9.3.

I have seen it happen a few times in schools where lots of children are rapidly affected and then it goes no further. It happened to a few classes at different times in Keyworth schools but more dramatically in a nearby Leicestershire village school in 2002 (Fig. 9.4). The headmistress of the Stathern primary school in the Vale of Belvoir contacted me because she suspected many pupils had whooping cough that she had diagnosed from www.whoopingcough.net (Chap. 14) and could not get any doctors interested.

These little patterns are of course lost in the national figures. Maybe a viral cold provides fertile ground and helps *Bordetella pertussis* to thrive so it can get a hold. That would mean the chance combination of a cold and pertussis would be a more severe infection than pertussis alone, but not only that, I think the combination

---

[3] Epithelium is the barrier layer on the surface of any kind of tissue. Our thick outside epithelium we call skin. The inside barrier layers are often delicate and thinner and sometimes have microscopic 'hairs' called cilia.

HEALTH SCARE: Whooping cough epidemic hits Vale school

40 PUPILS SENT HOME

WHOOPING cough has hit Stathern Primary School – and nearly half the pupils are off sick.

By ALICE RYAN

whooping cough vaccine

allowed the disease to spread. Mother-of-four Jackie Morris, of Blacksmiths End, said: "Children in the village

with suspected whooping cough.
"If it hadn't been for head-teacher Angela Wright push-

Fig. 9.4 Local newspaper reporting an event in 2002 that is very likely a daily event somewhere in the world (Nottingham Evening Post)

could increase infectivity considerably and explain the little 'mini-burst' of cases. It might also help to explain why it is so difficult sometimes to get a clear start date for the disease. Many times, a mother will say her child seemed to have a cold for weeks before starting to cough. I was able to look into this a little further at about this time.

Another of my attached medical students went through the medical records of families where at least two of the children had had whooping cough. She was look-ing for visits to the doctor by those children in the weeks before the whooping cough developed. She found that the first in the family to get whooping cough had had a prior consultation for a respiratory infection more often than the second or subsequent member. This again suggests *Bordetella pertussis* likes a previously damaged respiratory tract.

I have also found that the first case in a family is more severe than subsequent ones, for a similar reason I would speculate. One can imagine the probable mecha-nism here; if there is pre-existing damage to the respiratory tract and pertussis gets a hold, it will do more damage than if there were no previous damage. That would show itself as more severe symptoms.

Asthma is another example of predisposition. It is known that asthmatics are more susceptible to whooping cough, presumably because they already have inflam-mation in the lungs because of the asthma, and this therefore increases their suscep-tibility. Perhaps summertime 'hay fever' is the reason whooping cough outbreaks seem to start in the school summer holidays.

Several asthmatics who caught whooping cough told me that while they were suffering from it, and for a good while afterwards, their asthma was less trouble-some. I have no clue as to what might explain this, but it was volunteered sufficient times for me to believe it was a true phenomenon. No asthma sufferer ever told me it had made their condition worse.

*****

Once again I started preparing our findings for publication. It is often the hardest part of any project. Things never turn out exactly as expected and you have to make sense of your data and see how it might be presented in writing so that it fits into current knowledge by adding something important or new. The format of a scientific paper is similar the world over, starting with an introduction, then method, results, discussion and conclusion. It is usual to spend months on this, rewriting repeatedly to make sure it is clear to the reader. Part of the process is having it looked at by somebody who is an expert in the field for an objective criticism. If you can get it right before sending it to an editor, it increases the chance of acceptance enormously.

I had just discovered, to my great delight, that Dr. Ian Johnston had recently been appointed as a chest physician at the Queens Medical Centre, Nottingham. I knew Ian's name from several papers he had written about hospitalised patients in London with whooping cough, and a really important paper that looked for, and failed to find, evidence of persisting lung damage after whooping cough, [1] a question that many patients asked me up to this time, but I was only able to answer by saying I did not think so.

I was soon able to meet up with Ian Johnston after he gave a talk to GPs about asthma in the postgraduate centre at Nottingham City Hospital. I needed to pick his brains about another bit of research I was involved in that was troubling me and involved asthma. That research was focused on ways of teaching asthma patients to manage their own condition more effectively and I headed a four-person team investigating the effectiveness of two different educational tools, a book and a cassette tape. I was the lead researcher, Sue Jones was a health economist, Jan Davison was a statistician and Patricia 'Paddy' Hawtin was a nurse investigator who liaised with patients, along with my wife Joan, also a nurse, who was part of the team.

Paddy, in the course of discussions with the parents of these children with asthma in early 1985, had reported back to us that she was finding, quite by chance, that many of the children had been born by caesarean section, this being in an era when the section rate was about 1 in 25 in the UK. Fortunately we had the time and resources to delve more deeply into this, so having devised a protocol to look at all the birth factors we could think of, we obtained permission to access, thanks to the interest of Malcolm Symonds, the professor of obstetrics and gynaecology, the archived microfiche birth records of our asthma patients and an equal number of controls. Paddy, Joan and another nurse, Lynn Cator, went through all the traceable records thoroughly and painstakingly and we came up with a list of birth factors associated with our asthma patients. We were expecting to find a plausible relationship with some drug or procedure that at least made sense, but no, caesarean section was on its own the strongest associated factor. In the 124 asthma patients there were nine elective and four emergency sections and in the same number of controls there was one elective and two emergency sections.

This was startling. Everyone we told about it was intrigued but understandably highly sceptical. Their body language generally indicated, 'I can't believe it'.

Professor David Hull shrewdly said, 'Tiger country'. Respiratory medicine senior lecturer John Britton (later professor) did a sophisticated statistical analysis of our figures which resulted in a 'p' value of 0.11, not quite the 0.05 required to attempt publication, but we all agreed the next step was to repeat the process and use a larger sample. But as often happens, it got shelved because we had other preoccupations, and the team members moved on to other projects. This was 1989 and I think we were the first to spot the connection that has now been confirmed several times.[4]

*****

Ian Johnston was happy to look at the paper Jackie and I had written on the search for subclinical pertussis but I felt some apprehension as I had never met anyone else with an interest in whooping cough as comprehensive as his, and feared he would find some big flaw in our paper. Fortunately he thought it quite useful information and made several suggestions to make it clearer and more precise.

When I thought it was as good as we could get it, we sent it off to the *BMJ* with some optimism. They took a couple of weeks to reply with a 'No'. They had had it assessed by an expert in the field who decided the methodology was not precise enough to give a clear answer to the question we were asking. This was fair enough, but it was the only method we had, and it might not have given a clear answer but was still useful information we thought, so we sent it to another journal, *The Journal of the Royal College of General Practitioners* (JRCGP). As its name suggests this specialised in papers relevant to general practice, mainly from the UK. It was not as prestigious as the *BJM* but held in high regard and read by a large number of GPs. They accepted it on condition we shortened it (Fig. 9.5) [2]. That was not too difficult as it was a bit wordy. Unfortunately they shortened it even further by omitting the pattern of spread diagram. This was a shame as it was so revealing. Dr. Noel Preston of the Pertussis Reference Laboratory wrote to me about this afterwards as he had seen the diagram and thought it fascinating, as had Ian Johnston. Noel sympathetically pointed out that sometimes there was no fathoming editorial decisions and we all suffered from its effects from time to time.

Although this published study just covered 1985 cases, the outbreak continued right through 1986 and 1987 before settling down, both in Keyworth and nationally. We saw quite a few more adults with it in this outbreak, often caught from their children but by no means always. In these three years there were 105 cases, 30 of these being in 15-year-olds and older (29%), the start of a trend that was to continue. The generally accepted explanation was that more of the young ones were now being immunised so had more protection, and the generation born after 1952 when immunisation was introduced did not have the long-lasting immunity thought to occur after the natural infection.[5] Five adult patients said that they had had it before

---

[4] It is thought to result from the baby failing to pick up the mother's microbiome bacteria as in a natural birth. This can affect the baby's immune system which is how it relates to asthma which is an allergy-driven disease.

[5] The common belief that natural infection gave long-lasting immunity continued to distort perceptions for many years to come. It does not. It lasts roughly 15 years.

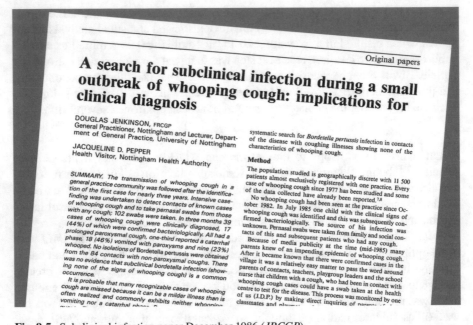

A search for subclinical infection during a small outbreak of whooping cough: implications for clinical diagnosis

Original papers

DOUGLAS JENKINSON, FRCGP
General Practitioner, Nottingham and Lecturer, Department of General Practice, University of Nottingham

JACQUELINE D. PEPPER
Health Visitor, Nottingham Health Authority

SUMMARY. The transmission of whooping cough in a general practice community was followed after the identification of the first case for nearly three years. Intensive case-finding was undertaken to detect contacts of known cases of whooping cough and to take pernasal swabs from those with any cough; 102 swabs were taken. In three months 39 cases of whooping cough were clinically diagnosed, 17 (44%) of which were confirmed bacteriologically. All had a prolonged paroxysmal cough, one-third reported a catarrhal phase, 18 (46%) vomited with paroxysms and nine (23%) whooped. No isolations of Bordetella pertussis were obtained from the 84 contacts with non-paroxysmal coughs. There was no evidence that subclinical bordetella infection (showing none of the signs of whooping cough) is a common occurrence.

It is probable that many recognizable cases of whooping cough are missed because it can be a milder illness than is often realized and commonly exhibits neither whooping, vomiting nor a catarrhal phase. ...

systematic search for *Bordetella pertussis* infection in contacts of the disease with coughing illnesses showing none of the characteristics of whooping cough.

Method

The population studied is geographically discrete with 11 500 patients almost exclusively registered with one practice. Every case of whooping cough since 1977 has been studied and some of the data collected have already been reported.[7,8]

No whooping cough had been seen at the practice since October 1982. In July 1985 one child with the clinical signs of whooping cough was identified and this was subsequently confirmed bacteriologically. The source of his infection was unknown. Pernasal swabs were taken from family and social contacts of this and subsequent patients who had any cough.

Because of media publicity at the time (mid-1985) many parents knew of an impending epidemic of whooping cough. After it became known that there were confirmed cases in the village it was a relatively easy matter to pass the word around parents of contacts, teachers, playgroup leaders and the school nurse that children with a cough, who had been in contact with whooping cough cases could have a swab taken at the health centre to test for the disease. This process was monitored by one of us (J.D.P.) by making direct inquiries of parents ...classmates and play...

Fig. 9.5 Subclinical infection paper December 1986 (*JRCGP*)

as a child but I was unable to verify this. I had also now seen a child 3yf[185] who had had it twice, once in 1977 and again in 1987 13yf[431].

The next surge we could expect would start in 1989 after another two years. In the meantime, I had plans to extract more information out of the cases I had studied so far.

References

1. Johnston IDA, Anderson HR, Lambert HP, Patel S. Respiratory morbidity and lung function after whooping cough. Lancet. 1983;322:1104–8.
2. Jenkinson D, Pepper JD. A search for subclinical infection during a small outbreak of whooping cough: implications for clinical diagnosis. The Journal of the Royal College of General Practitioners. 1986;36(293):547.

Chapter 10
1987. Does the Effectiveness of the Vaccine Wear Off?

One of the unexpected findings of my first study in 1977–78 was that out of 141 one-year-olds, 24 had caught whooping cough and they were all unimmunised (Fig. 10.1). None of the immunised ones, 43 of them, had caught it. In the two- to four-year-olds a small number of the immunised caught it. On the face of it the one-year-olds seemed to get complete protection. This contrast begged the question of why only the one-year-olds seemed so well protected, but from a scientific viewpoint there was substantial doubt about whether it was telling the real story. In the first place the numbers were relatively small by epidemiological standards and the absence of any in the immunised group might have happened because of some unappreciated bias. There might also have been unmeasurable human factors operating over that short time. For instance, you could argue that mothers who had their babies immunised were more scared of whooping cough, so when it arrived in the village those same mothers kept their one-year-olds at home out of contact with the disease to protect them, whereas the mothers who did not get their child immunised were not so scared and allowed more normal mixing. It is easier to keep a one-year-old isolated than a two- to four-year-old! One cannot say that really happened, but it might have happened, and it is impossible to prove either way, so it is a possible confounding factor and casts doubt on the apparently clear result. This is an example of how scientists are sceptical. Trying to find fault with results of investigations is the way to get as close to the truth as possible.

By 1987 I had been collecting cases for ten years and had well over 400. All this time, one of the things I kept thinking about was, if the effect of the vaccine wore off with time it might explain why we still kept seeing so many cases in the immunised. The UK was somewhat unusual insofar as only three doses were given. Other countries gave a fourth booster dose later on, which implied some possibility of the early benefit wearing off. Some researchers had tried working out the protection at different ages by looking at attack rates at different ages when studying it on much larger groups than I had. There was sometimes a suggestion that it wore off but there were so few studies like that with no consistency between them so the whole

D. Jenkinson, *Outbreak in the Village*, Springer Biographies, https://doi.org/10.1007/978-3-030-45485-2_10

Year of birth	1973		1974		1975		1976		1977	Total
Whooping cough Immunisation	+	−	+	−	+	−	+	−		
Children affected by whooping cough	9	17	7	35	2	23	0	24	9	126
Children not affected by whooping cough	96	18	84	41	74	39	43	74	78	547
Vaccine Protection	90%		90.2%		96.4%		100%			

Fig. 10.1 Whooping cough attack rates in 1977–78 as published in the *BMJ* 'Outbreak' paper in 1978 (where the 'Vaccine protection' row was actually the 'Odds ratio') showing 100% protection in 1-year-olds

question was still wide open. But as the years wore on and I never saw whooping cough in an immunised one-year-old ever, I became all the more convinced that full individual protection was only complete for about 12 months.

All the previous studies were cross-sectional, that is to say they made their observations at a single point in time looking at different ages at the same time, so there might have been important unknown differences between age groups that affected the result. I did not have sufficient numbers affected in each outbreak for the numbers to reach adequate statistical significance,[1] but I wondered if somehow I could lump all the data together for a result in what would be a longitudinal study, that is, following the same group of people over a prolonged period of time.

I assumed somebody would have developed a method of combining the numbers in a cohort of patients to calculate age-specific vaccine protection, but I could not find anything remotely similar or even referring to such a method in the library indexes. I asked all the experts I knew in the medical school and they were unable to help. I just had to put the idea on one side hoping I would eventually come across something.

I cannot remember how the idea originally came to me but at some point I realised that I was not going to find the solution in print, so perhaps I should set about working a method out for myself. Computerised spreadsheets were just starting to become available and it seemed as if such a thing might provide a way, but in the end I just sat down with a pen and paper and eventually it came to me. I am not mathematically inclined so could not see the solution as a whole. I worked each

---

[1] Significance level has a precise meaning. In the biological sciences, observations are generally only taken seriously if there is a less than 5% possibility that those observations could have happened by chance. There is a whole branch of mathematics devoted to the subject.

little step at a time until I came to realise it was essentially quite simple but only if I made a big assumption. I needed to assume that the population had not changed its overall numbers in each age group, or in the proportion that was immunised or in the proportion that already had whooping cough. Clearly these things must have changed at least a little over the eight years my calculations were to apply to, but if the changes were small then the result could still be valid. If the changes were big it could render the result inaccurate. First, I tried to measure the possible changes that had occurred.

I worked out that the overall population of the practice changed at 7% per year, but families with children had a much lower rate. Over three quarters of those that had whooping cough in 1977–78 were still registered ten years later, and the overall practice population had only grown by 3% in eight years. I was unable to estimate if those coming into the practice had had whooping cough, but I thought it reasonable to assume it was the same proportion as those leaving. The proportion immunised was definitely likely to be different, as our rate was generally higher than elsewhere. None of these potential inaccuracies should prevent a useful result I thought.

I will explain the method step by step. It is actually quite simple. Let's start with all the one-year-olds born in 1976[2] and divide them into four groups, immunised and unimmunised and affected and unaffected by whooping cough when they were between one year old and two years old (Table 10.1).

Twenty-four unimmunised had whooping cough. We now look at the same children when they are two years old (over 2 but under 3) but because 24 have had whooping cough they are now assumed immune and deducted from the unaffected group because we are only interested in those still not immune. 92 minus 24 is 68. So 68 now move into the two-year-old boxes (Table 10.2).

Still looking at those born in 1976, three unimmunised caught whooping cough when they were two so 68 minus three is 65. So 65 would move right, into the three-year-old box that is not actually shown in Table 10.2. The same process repeats with each yearly increase to the right and stops at seven years old where in this cohort there are still 42 in the 'immunised and no whooping cough' box, but only 52 left in the 'unimmunised and whooping cough' box (Fig. 10.2). In this particular birth year

Table 10.1 1976 births. Affected and unaffected and immunised or unimmunised while over one but under two years of age

Born 1976		While 1 year old	
		Whooping cough	No whooping cough
	Immunised	0	42
	Unimmunised	24	92

---

[2] The eagle eyed will notice that the numbers are different from figure 10.1 which was the 1978 count. The table 10.1 figures are based on our 1987 register, as it was not possible to recreate the 1978 one. This has distorted the unimmunised number.

Table 10.2 1976 births affected at one and two years old

Born 1976		While 1 year old		While 2 years old	
		Whooping cough	No whooping cough	Whooping cough	No whooping cough
	Immmunised	0	42	0	42
	Unimmunised	24	92	3	68

Table 10.3 1976 and 1977 births affected at one and two years old

Born 1976		While 1 year old		While 2 years old	
		Whooping cough	No whooping cough	Whooping cough	No whooping cough
	Immunised	0	42	0	42
	Unimmunised	24	92	3	68
Born 1977		While 1 year old		While 2 years old	
		Whooping cough	No whooping cough	Whooping cough	No whooping cough
	Immunised	0	42	0	42
	Unimmunised	2	53	0	51

none of the immunised caught whooping cough right up to the age of seven. Perhaps the batches of vaccine were particularly good that year! It is quite possible.

The whole table is now seven years wide for the 1976 cohort, but we need to add other birth years. These are added above or below according to the year of birth. I will illustrate how by adding the 1977 cohort which starts immediately below Table 10.2 thus Table 10.3.

Only 2 of those born in 1977 caught whooping cough when a year old. They were both unimmunised so at age two it was the number 53 minus 2 that became the number of susceptibles.

The table became quite long as rows for every available year when there were data were added. There were 16 rows in all from 1970 to 1986.

When the table is complete all the columns are added up, so you end up with a single row divided into immunised and unimmunised and seven columns for each year one to seven.[3] The vaccine protection rate is then calculated as explained in chapter five. The full table is shown in Fig. 10.2 in a more concise form, the way it was published in the *BMJ*.

Table 10.4 shows the age-specific vaccine protection rate.

The paper was fairly straightforward to write up as it was an entirely novel method of calculation whose justification was quite impossible for me as a

---

[3] Age 1 means between 1 and 2, the second year of life etc. Under ones are ignored because immunisation is not complete until 11 months and it is not considered effective until some weeks after the final third dose.

Year of birth	Immunisation against pertussis	Age (years) 1	2	3	4	5	6	7
1970	+							9 (210)
	−							0 (10)
1971	+					2 (188)	10 (186)	1 (176)
	−					0 (10)	1 (10)	0 (9)
1972	+				1 (189)	20 (188)	0 (168)	0 (168)
	−				0 (10)	4 (10)	0 (6)	0 (6)
1973	+			1 (114)	12 (113)	1 (101)	0 (100)	0 (100)
	−			4 (64)	12 (60)	0 (48)	0 (48)	1 (48)
1974	+		1 (95)	6 (94)	1 (88)	0 (87)	0 (87)	0 (87)
	−		0 (84)	36 (84)	8 (48)	2 (40)	1 (38)	1 (37)
1975	+	0 (85)	1 (85)	1 (84)	0 (83)	1 (83)	3 (82)	1 (79)
	−	1 (70)	21 (69)	5 (48)	1 (43)	1 (42)	4 (41)	0 (37)
1976	+	0 (42)	0 (42)	0 (42)	0 (42)	0 (42)	0 (42)	0 (42)
	−	24 (92)	3 (68)	7 (65)	1 (58)	3 (57)	2 (54)	0 (52)
1977	+	0 (42)	0 (42)	0 (42)	1 (42)	0 (41)	0 (41)	0 (41)
	−	2 (53)	0 (51)	3 (51)	5 (48)	3 (43)	0 (40)	0 (40)
1978	+	0 (83)	0 (83)	1 (83)	0 (82)	0 (82)	5 (82)	1 (77)
	−	2 (54)	3 (52)	6 (49)	3 (43)	0 (40)	4 (40)	3 (36)
1979	+	0 (79)	0 (79)	0 (79)	0 (79)	1 (79)	2 (78)	2 (76)
	−	1 (52)	3 (51)	1 (48)	0 (47)	0 (47)	4 (47)	3 (43)
1980	+	0 (94)	0 (94)	0 (94)	0 (94)	3 (94)	4 (91)	1 (87)
	−	3 (20)	1 (17)	0 (16)	0 (16)	3 (16)	3 (13)	0 (10)
1981	+	0 (95)	0 (95)	3 (95)	1 (92)	1 (91)	3 (90)	
	−	0 (20)	0 (20)	0 (20)	6 (20)	3 (14)	1 (11)	
1982	+	0 (113)	1 (113)	2 (112)	0 (110)	0 (110)		
	−	0 (17)	0 (17)	0 (17)	2 (17)	0 (15)		
1983	+	0 (83)	0 (83)	0 (83)	0 (83)			
	−	1 (10)	2 (9)	2 (7)	0 (5)			
1984	+	0 (98)	0 (98)	2 (98)				
	−	0 (9)	0 (9)	3 (9)				
1985	+	0 (81)	0 (81)					
	−	0 (19)	0 (19)					
1986	+	0 (91)						
	−	1 (6)						
Total	+	0 (986)	3 (990)	16 (1020)	16 (1097)	29 (1186)	27 (1047)	15 (1143)
	−	35 (422)	33 (466)	67 (478)	38 (415)	19 (382)	20 (348)	8 (328)
Attack rate	+	0	0·003	0·016	0·015	0·024	0·026	0·013
	−	0·083	0·071	0·140	0·092	0·050	0·057	0·024

Fig. 10.2 Cumulative attack rates in 1- to 7-year-olds. Number of immunised and unimmunised affected in each year with unaffected numbers in brackets (*BMJ*)

Table 10.4 Age-specific vaccine protection rates

1y	2y	3y	4y	5y	6y	7y
100%	96%	89%	84%	52%	54%	46%

non-mathematician to even start. I just had to leave it to the mercy of the assessors. The numbers all had to be carefully checked of course, and the table of results looked a bit of a nightmare for printing.

It was accepted by the *BMJ* and was published on 27th February 1988.

It is sometimes acceptable in a paper to make a recommendation as part of a conclusion to it if the findings point to a particular deficiency in current treatment practices for instance. I suggested that perhaps a pertussis element should be added

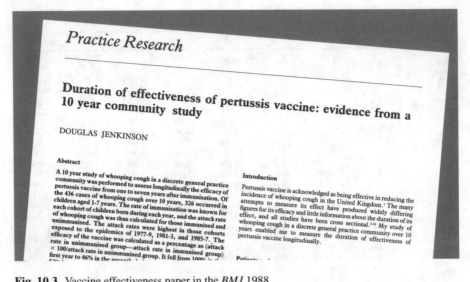

Fig. 10.3 Vaccine effectiveness paper in the *BMJ* 1988

to the pre-school booster injection that was at that time just diphtheria and tetanus with polio drops by mouth. I think there were quite a few people who would have agreed this to be a good idea, but the brain damage controversy was still fresh in people's minds and it would have taken a very brave government to put a further dose of the old vaccine into children, particularly when they knew that purer and less reactive vaccines were being developed.[4]

Of all my papers on whooping cough this is the one most often cited in other papers, but there was another spin off from it that I was not expecting. Much later in 1996 I was fortunate to meet the foremost UK researcher into pertussis issues, Professor Elizabeth Miller, who was head of the Immunisation Division of the Public Health Laboratory Service (*PHLS*) Communicable Diseases Surveillance Centre, at Colindale, London. I knew of her and her eminent reputation in the field already. She and Douglas Fleming, the director of the Manchester *RCGP* research unit, were the two external examiners for my Doctor of Medicine degree viva. David Hull was the internal examiner.

In mid-viva Elizabeth Miller said, 'Did you know that papers had been written about your 'Duration of effectiveness' paper?'

'No', I said, suddenly feeling worried that she had exposed an enormous chasm in my knowledge that I would probably now get buried in, and immediately thinking the only reason anyone would do such an unusual thing as write a paper about a paper would be to 'rubbish' it. She explained that a mathematician in her department, Paddy Farrington, had analysed my figures in order to calculate the possible errors inherent in the method, so that others could use a similar method but express the results in a more scientifically accurate way. This was an era when medical

---

[4]Acellular pertussis vaccine was eventually added to the pre-school booster in 2001.

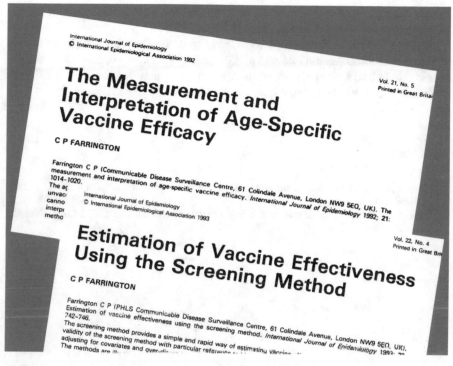

Fig. 10.4 CP (Paddy) Farrington's 1992 and 1993 papers about my 'duration' paper (*International Journal of Epidemiology*). Reproduced by permission of Oxford University Press

researchers were needing to embrace the skills of medical statisticians in order to meet the demands for an evidence base for any proposed advance, and medical journals were increasingly using statisticians on their editorial panels. I think up to this point my papers had just crept in under the detailed statistical analysis radar.

Paddy Farrington's papers were way beyond my understanding and full of mathematical symbols and formulae that were most impressive [1, 2]. The bottom line was that so many assumptions were made; the potential for error was enormous in mathematical terms. I might have been right, or completely wrong, depending on whether the assumptions were correct or not. I knew I had made several assumptions, but he managed to find many others I had not even thought about. I now tell myself he was right. But so also, as it turned out, was I!

Some of my misconceptions about vaccines had been pointed out to me in the 1970s when I naively thought it was doctors battling infectious diseases who decided the government's immunisation policy. I could not have been more wrong. Diseases spread in a way that can be expressed mathematically quite accurately in many cases, and likewise the effect of a vaccine. The two can be reconciled in complex mathematical formulae that tell the best strategy for vaccine deployment. Nowadays this kind of modelling is accomplished with the aid of computers like weather forecasts, but what it amounts to is that immunisation policy is decided by mathematicians in back rooms who don't need to have the slightest clue about what diphtheria or measles or whooping cough ever look like.

Back in 1988 it seemed things were fairly quiet on the whooping cough front. The national immunisation rate was at 75% and still creeping up, and the next epidemic due in 1989 was only going to turn out to be half the size of the 1985–87 outbreak, but by its end I would have notched up my 500th case.

References

1. Farrington CP. The measurement and interpretation of age-specific vaccine efficacy. Int J Epidemiol. 1992;21:1014–20.
2. Farrington CP. Estimation of vaccine effectiveness using the screening method. Int J Epidemiol. 1993;22:742–6.

Chapter 11
1988. Trying to Raise the Uptake of Pertussis Vaccine

The record low uptake for pertussis vaccine of 31% for 1978-born children had climbed back to 73% by 1987 partly as a result of persuasion of parents by doctors and health visitors that the vaccine was safe and that whooping cough was dangerous, but mainly, I suspect, because the media had stopped sensationalising any remaining controversy, thus removing the main conflict of trust. A target of 95% was considered reachable and many local authorities were taking it upon themselves to undertake local promotional activities in medical centres and public places with the help of midwives, nurses, health visitors and doctors, often in their free time. The Department of Health was usually behind these regional initiatives.

Considerable experience had been gained with a 'Mop up Measles' campaign in Nottingham some years previously to promote the uptake of measles vaccine, so a new working party was set up by the Nottingham Health Authority to organise and co-ordinate local activities promoting pertussis vaccine. I found myself on it having dreamt up the measles slogan and worked enthusiastically on that initiative. This campaign used the not so punchy slogan, 'Keep whooping cough out' on posters and display stands throughout the area and in other publicity activities such as local radio slots and interactive exhibitions in shopping centres.

One of the biggest problems that was slowing down the achievement of a high uptake level was the misconceptions among medical staff of the contraindications to giving the vaccine. Since the 1974 brain damage scare, which was now proved virtually baseless, doctors (who were the only professionals at that time able to administer immunisations) were often still erroneously excluding children with certain medical histories, or with relatives with histories, because of confusing guidance.

It was understandable because there were differences between what experts were saying and what was stated on the information sheets that came with the vaccines, and even these would differ between manufacturers. Manufacturers tended to use phrases like 'caution should be exercised' when giving it to patients with certain medical conditions, but many doctors would interpret 'caution' as 'do not' and deal with the ambiguity by playing safe.

D. Jenkinson, *Outbreak in the Village*, Springer Biographies,
https://doi.org/10.1007/978-3-030-45485-2_11

In such circumstances a trusted, authoritative, reasoned and justified directive can give doctors (who still take personal responsibility for their actions) the confidence and support needed to be more positive.

On the pertussis working party I met Angus Nicoll, then a lecturer in child health. He had a great talent for understanding how patients and medical staff thought and how change might be brought about. He later became a professor of epidemiology and expert on influenza with an international reputation.

Angus and I compared notes and realised that overcautious doctors were a big issue, and that even if they were confident themselves, they could be undermined by well-meaning and trusted health visitors who were themselves being overcautious and might already have sowed seeds of doubt in parents' minds that were difficult to reverse.

We decided there were several ways we could help improve the uptake of pertussis immunisation. One was by encouraging 'catch-up' pertussis immunisation for those who had missed out, by providing monovalent pertussis vaccine[1] which would need to be given in three doses. This would enable those who had decided against it for the primary immunisation and now wanted to put the matter right. Angus would also hold training sessions to update the staff who were doing the immunisation counselling.

We also thought it would be a good idea to produce an algorithm (decision chart) showing the decision issues relating to the administration of triple vaccine (*DTP*), but crucially, we wanted to give users some quantitative measure of the probability of going down any specific branch. We thought this would be of real help to someone unfamiliar or lacking confidence. To achieve this we set about counting the numbers going down each of the different branches at the various decision points over the course of many immunisation clinics as we applied 'best practice'.

To make it even more useful we included the mumps, measles and rubella (*MMR*) immunisation too, because both tasks are usually undertaken concurrently. It was published in the *BMJ* in August 1988 (Fig. 11.1) [1].

In addition we listed the false contraindications that some health workers erroneously believed meant that pertussis vaccine should be omitted (Fig. 11.2). They had probably arisen out of an instinct to give it only to strong and healthy babies if it was possibly hazardous. None was evidence based; indeed, the opposite could be argued, that it was babies with problems that were most in need of the protection it gave.

The probability guidance is not something I have seen repeated in similar charts although to my mind it is invaluable. Perhaps I am just unaware of others. Maybe it is not common because in many instances the information to base figures on has not been collected (Fig. 11.3).

It is impossible to know the impact of these activities as they were a small part of a continuous national drive but in the next five years the pertussis vaccine acceptance rate increased from 73% to 92% and reached a record 96% in 2010.

---

[1] Containing pertussis vaccine only. Sadly, no manufacturer now produces this potentially useful monovalent pertussis vaccine.

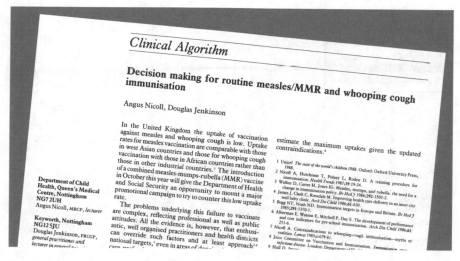

Fig. 11.1   Algorithm 'with a difference' (showing proportions). *BMJ* 1988

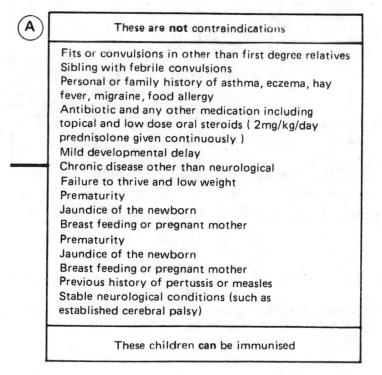

Fig. 11.2   Falsely believed contraindications that sometimes unnecessarily prevented children getting pertussis vaccine. *BMJ* 1988

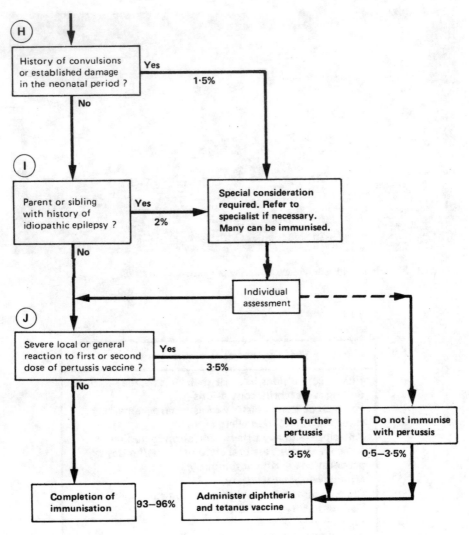

Fig. 11.3 Part of the decision tree with actual probabilities marked at branches. *BMJ* 1988

Reference

1. Nicoll A, Jenkinson D. Decision making for routine measles/MMR and whooping cough immunisation. BMJ: British Medical Journal. 1988;297(6645):405.

Chapter 12
1989–91. The Fourth Outbreak.
The Natural History of Whooping Cough

When anyone starts talking about disease it is usually about how that disease affects them or somebody they know. They don't normally talk about the disease itself. A full description is how you might expect medical professionals to be taught or something you might discover on Wikipedia. How a disease affects people and its outcome without medical intervention is the natural history of a disease. It is the foundation knowledge of medical treatment because the natural history is something we usually want to change if we can because it is painful or serious in some way.

Learning about a disease normally happens in a structured way. We often start with the cause: in the case of whooping cough *Bordetella pertussis*. Then what the sufferer is likely to complain of: the symptoms, such as coughing, vomiting and a feeling of suffocation. Then the things a doctor might find on examination: the signs, none in most cases of whooping cough but occasionally bruising in the white of the eye or micro-bruises in the skin of the face, or rarely a fractured rib. Then the outcome, which is what will happen if you do nothing: usually full recovery in the case of whooping cough. Then possible treatments: usually none in uncomplicated whooping cough except infants in hospital. There are other aspects too, such as how the body is damaged (pathology), how a professional manages the illness or helps others to manage it, such as antibiotics to stop spread, ensuring parents sleep near a youngster with whooping cough, what laboratory tests might be helpful, and quite a few others.

Considering the whole of the natural history of a disease as a story with a beginning, a middle and several possible endings is worthwhile because normally we only look at or consider part of a disease and its effects on us or our friends and relatives, and that is very restricting when judgements need to be made about taking advantage of immunisation for example.

One of the most challenging aspects of a GP's job is needing to know something about every disease under the sun, something more about every kind of disease one's patients might get, and a lot about the diseases they commonly get. Spending

© The Editor(s) (if applicable) and The Author(s), under exclusive license to
Springer Nature Switzerland AG 2020
D. Jenkinson, *Outbreak in the Village*, Springer Biographies,
https://doi.org/10.1007/978-3-030-45485-2_12

five years at medical school, then one or two years of pre-registration in hospital jobs, followed by three years of GP specialist training gives one a fair start in the knowledge stakes, but you never ever stop learning because individuals differ in the symptoms they get with the same disease. Treatments change too, so you have to keep up with that. The half-life of medical knowledge is about five years supposedly, so of all the things you knew 10 years ago, only a quarter are still correct. You would have to be superhuman to keep on top of it all, so you can't. Instead you have to be very aware of the limitations of your own knowledge and keep an open mind, re-educate yourself continually, and always be ready to refer on to somebody more senior. Being sure about something is possible in medicine but fraught with danger.

Notwithstanding all this, the UK GP is in an advantageous position when it comes to providing a comprehensive service. Patients can only be registered with one practice, so it is quite clear who is responsible for care, and there is a single record of it all. Although there are some disadvantages, it seems better than places like the USA where patients can shop around, and doctors compete with each other to make a living. I have always told people to beware of any doctor who has a vested interest in your illness.

In the UK the NHS leans the other way. Because almost everyone in it works hard and employment is secure, *their* vested interest is in making you well and keeping you that way, which is how it should be, and its pyramidal structure ensures that difficult problems can go higher up the pyramid if necessary, to find a solution. The downside is that it is so effective it leads to rationing by waiting, but the main alternative, which is to ration by ability to pay, would be intolerably worse.

This NHS ethos has fostered altruism in its employees, directed not only at patients but at the service in general through mutual cooperation and service development. These are things that can get stifled in commercial organisations where such activity might give advantage to competitors. Thus, many employees at all levels since the NHS started have done research freely alongside their daily tasks when they have felt the inclination, and generally had encouragement from management and peers.

When I first worked in a general practice environment during a holiday from hospital in 1968, I did some GP locum work for Dr. Anderson in Mablethorpe, Lincolnshire, and I felt hopelessly out of touch with the problems patients were bringing. The only knowledge I had was what I had learned in five years at medical school and a year in hospital general medicine and surgery wards. I knew little or nothing about how to manage the apparently 'minor illnesses'[1] patients mainly turned up with. I think I was probably able to do the one thing that is essential, but very difficult to get right. It requires a doctor to be able to distinguish between somebody who is seriously ill and someone who isn't. Get it wrong just once and you run the risk of being struck off the medical register and probably ending up in

---

[1] Minor illness might not sound important until you understand that most major illnesses start as something apparently minor.

court as well. I had a lot to learn, and this was before organised GP training was available, and even longer before it was obligatory.

By the time I found myself as a proper GP in 1974, help had arrived in the form of two very recently published books, Keith Hodgkin's *Towards Earlier Diagnosis* [1] (1973) and John Fry's *Common Diseases* [2] (1974). These two GPs in very different parts of England, Redcar and Beckenham, respectively had found themselves with the same problem as me in Mablethorpe but several decades earlier. They had also found themselves in unfamiliar disease territory and seriously medically disorientated outside hospital. Out of thousands of GPs struggling with the same issues, these two doctors set about systematically recording the multitude of consultations that came to them every day, classifying and counting symptoms, diseases and problems, their frequency and outcomes over many years. These books became the bibles for doctors training for general practice or already in it, and almost at a stroke defined what the discipline of general practice was about. Until about then, general practice was not considered a specialty at all but rather as what you did if you 'Fell off the specialist ladder', to quote Lord Moran, who was Winston Churchill's doctor. General medical practice is a relatively recent discipline that became defined and shaped in the orbit of the new NHS after 1948. The first professorship of general practice in the world was only established in 1963 in Edinburgh.

What Fry and Hodgkin had actually done was describe the natural history of common problems presenting to GPs. It is very important for any doctor to know the natural history of the illness being treated and it should also be important for patients who are suffering from it. Otherwise how will they know if a doctor is necessary at all.

In spite of the work of Hodgkin and Fry who did it all without computers, we still do not have this information for very many diseases, even though so much is recorded on computer. It is true there are excellent data on subsets of patients with certain diseases in certain situations, but for the majority there is little, because the information cannot be recorded in a meaningful way unless the computer demands it, and because computers do not care about accuracy any more than a written record does, the maxim 'rubbish in, rubbish out', can still apply.

To give just one example. Many people develop shingles.[2] Most probably go to a doctor if they get it. It will almost certainly be recorded on the computer with the date of diagnosis. In order to know the natural history, you need to know when it started and when it finished, the duration of the rash, and the neuralgia, and the same for the debility. In all probability none of that will be recorded except perhaps occasionally in free text which is usually not analysable anyway.

To know the full natural history you need to know the variation in symptoms in every single one of the patients who present with it. How many get it mildly? How many get it severely? How many have permanent disfigurement or disability? Then

---

[2] Shingles is a localised rash, often painful, in the area of skin served by a nerve branch. It is caused by reactivation of the chickenpox virus that has stayed in the body.

you need to know, in addition, how things might be different with different forms of treatment.

If you want the answer today, I suspect you still might refer to Hodgkin or Fry. You can also look up specialist papers or books, but they are likely to have the same basic weakness. They do not start with *all* the people who are getting it, just a selected group. Even textbooks are usually unable to quantify the full range of the natural history. And that is just shingles!

It might change when we can interact with our own medical records and put all the information in for ourselves. And that may not be far away.

*****

I started thinking about the possibility of describing the natural history of whooping cough at the end of the small fourth outbreak which started in 1989.

The 1989–91 outbreak of whooping cough turned out to be only half the size of that of 1985–87 with a total of 55 cases in Keyworth (Fig. 12.1). This could be attributed to the increased proportion of immunised in the most susceptible age groups. The average age of those affected was higher than in previous outbreaks for the same reason. It was becoming quite common to see it in teenagers too, and I had a stark reminder of how susceptible babies are.

Baby girl, f1m[498], was born on 4th April 1991. She had three older siblings, all of whom had been immunised. The three-year-old sister f3y[499] started coughing on 15th April but it was not a bad cough. There was no vomiting or apnoea, just relatively mild paroxysms, and she had recovered from it five weeks later, but it was definitely whooping cough. The two-week-old baby sister had no chance of escaping it, being exposed at the most susceptible age to the adoring three-year-old sister with paroxysms of coughing.

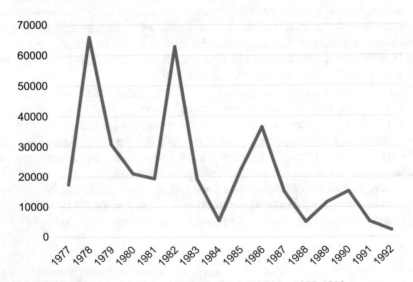

Fig. 12.1 Notifications of whooping cough in England and Wales 1977–1992

After being at home just three weeks, on 26th April 1991 the baby started to become ill with a cough. It slowly built in severity and after a week or so she was vomiting and going blue after every paroxysm of coughing. It was 16th May before medical attention was sought and I was able to examine her and take a per-nasal swab.[3]

The next day she had deteriorated badly and had to be admitted to hospital with pneumonia. The swab came back positive for *Bordetella pertussis* five days later, and so the cause was then clear and we all knew what we were dealing with, but it was only the day before admission, when I talked to the mother and took the swab, that I discovered the three-year-old sister with whooping cough. The older sister of five and brother of six seemed to have escaped, but in fact, the five-year-old f5y[500] was incubating it and went down with it 10 days later!

It all went to show how unpredictable a disease it is. Middle sister gave it to baby sister but not to older sister. The baby could not stop middle sister coughing over her, but perhaps older sister went away from her when she coughed. Older sister developed it eventually, several weeks later, presumably when the baby was cough-ing so badly that the whole family would be taking turns to nurse her. The six-year-old brother escaped as did mum and dad. The baby made a good recovery eventually.

Big sister was case number 500. I probably just smiled to myself at the time but later the idea that I should develop something from a landmark number strength-ened. Another aspect that gave me food for thought was the fact that I did not see another case for over 12 months, so the combination of that number and a long interval seemed all the more significant. Perhaps that was the end of whooping cough! Perhaps we had immunised it out. Perhaps it would not come back again, ever!

Although I was aware at the time that I was seeing and diagnosing more whoop-ing cough than most doctors, it did not really occur to me that I might be able to change that situation. I just felt it was the natural consequence of taking an interest. The information I had gathered was to do with the effectiveness of immunisation and nothing else. The lull in cases gave me time to think about other ways of looking at the data and I eventually wondered if I had enough to write an account of the natural history of whooping cough. I certainly had some information, but would it translate into new knowledge that might be useful? I knew it would involve a great deal of work, but it might be worth it. I would give it a go. I thought hard about the best way of going about it and in the end decided writing a specific paper about the natural history of whooping cough might be best. As far as I was aware this had not been done on a general practice population in the UK before, apart from the works of Fry and Hodgkin, which were not detailed.

To do this kind of analysis would normally involve retrieving all the NHS records to extract the information, something which is extremely difficult when records move around the country with the patient. By recording all 500 cases on separate

---

[3] This scenario is now avoidable. Since the introduction of a pregnancy booster injection for pertus-sis in 2012, there has been a dramatic drop in the incidence of whooping cough in babies of immunised mothers.

sheets of paper I had avoided that problem and an even bigger one by having structured the records in a way that gathered the data required to analyse natural history, even though it was not the original intention. Unfortunately, 500 was just too big a number of sheets of paper to analyse by juggling them round the living room floor. Newly appearing personal computers came to my rescue.

Computers have always fascinated me. Perhaps it is to do with curiosity about how things work. I thought the Sinclair ZX81 was the best thing ever, and I remember paying £99 for one in 1982. There was nothing useful to be done on it, but it had one of the clearest programming manuals ever written, so I learned to program in BASIC.

The next step up was to a BBC B computer, very versatile for the time but with few useful programs and very limited storage. It could handle twin double-sided floppy disks at 640kB each. Fortuitously by 1991 when I needed decent computer power to analyse the 500 cases, affordable cloned IBM 8086 machines were available that were enormously better.

The first requirement was a database program into which I would transfer my handwritten information. The godsend that came out in 1985 which I did not discover until a couple of years later was EPI-INFO, a public domain database program that could be configured easily to individual requirements. Created by the Centers for Disease Control and Prevention (*CDC*) in Atlanta, Georgia, it was meant to facilitate scientific study in third world countries and help investigators without financial resources. It would run on simple and old computers under MS-DOS[4] and was only changed in 2000 to something similar to Microsoft Excel. It is still going strong, and even the original DOS version, the last incarnation of which was in 2001, is still running in a few places and is still supported!

The other requirement was a statistics program and I found 'Easistat' ideal. That was a MS-DOS program too and I believe in the public domain, but it has disappeared from view today. The modern EPI-INFO I believe does it all.

So that was my evenings sorted for a good few months as I fed the PC with data.

The 500th case was recorded in May 1991. The final draft of the completed paper was in December 1993 and it was eventually accepted by the *BMJ* in September 1994 after some valuable modification suggested by Ian Johnston and some shortening that the journal wanted.

I had spent two years working on the figures. It is all a bit of a blur now as it was just slotted into the bits of free time I could find between my general GP work and having a busy family life. I am only reminded of the magnitude of the task by a thick folder of printouts of numbers by Easistat, most of them with my notes and calculations cryptically scrawled on them. I didn't just have to provide the raw numbers but also the 'p' values and confidence intervals too, as the standard required by journals got higher and higher.

---

[4] MS-DOS was the operating system developed by Bill Gates and his company Microsoft that came to dominate PCs.

I am not sure why the *BMJ* changed the title from 'natural history' to 'natural course' but I suspect they thought it would make more sense to readers for whom English was not their first language. There has always been a large number of them. It is also probably a better search term.

The published paper came out in February 1995 (Fig. 12.2) [3] and is packed with information but not in an easily digestible form. It is basically a list of facts and some opinions. Here is a list of the facts.

- The interval between attacks of paroxysmal coughing varied from 15 minutes to 12 hours.
- The average number of paroxysms was 13.5 per 24 hours.
- The change from an ordinary cough to a paroxysmal cough ranged from 3 to 14 days.
- 6 of the 50 adults affected recalled having it as a child.
- One child had it at 3 and also at 13.
- Many children had only 3 or 4 paroxysms per 24 hours, mainly at night.
- In childhood it was commonest in three-year-olds.
- In adulthood it was commonest in the fourth decade.
- 219 were male. 281 were female.
- In adults 16 were male. 34 were female.
- The average duration of the cough was 52 days (range 2–164).
- The younger you were the longer the cough lasted.
- The greater the frequency of paroxysms the longer the cough lasted.
- Only a third of patients had initial catarrhal symptoms. (Based on a sample of the 500.)

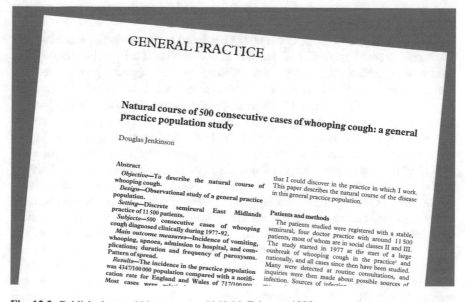

Fig. 12.2   Published natural history paper *BMJ* 4th February 1995

- 57% vomited at some stage with paroxysms.
- 49% whooped at some stage.
- 11% had apnoeic episodes.
- The frequency of vomiting, whooping and apnoea was the same in adults and children but greater in females than males.
- Females had more frequent paroxysms (15.1 v. 11.5 per 24 hours).
- The immunised had less vomiting (49% v. 65%).
- The immunised had less whooping (39% v. 56%).
- The immunised had less apnoea (8% v. 15%).
- The duration of coughing was less in the immunised (49 days v. 55 days).
- The number of paroxysms was less in the immunised (11 v. 15).
- Positive cultures were obtained from 25% of the immunised v. 52% of the unimmunised.
- 5 patients had serious complications (pneumonia).
- The incidence was 6 times greater than the notification rate in England and Wales.

None of these findings would be any great surprise to most people but there is a big difference between believing something to be true and actually knowing it is true and being able to put precise figures onto it. Most people had thought being immunised meant you had only a mild attack if you actually succumbed to it, but I was a little surprised to find it only made a relatively small difference. But on the other hand, if you were immunised there was only half the chance of finding the bacteria compared with if you were unimmunised. This implies that you are less likely to pass it on and is direct evidence supporting the importance of herd immunity in the spread of *Bordetella pertussis*. It confirms what we were observing already from the notification figures that increasing the immunisation rate by say 20% might cut the number of cases by 40% (this is a total guess on my part for illustration of my point).

One of the biggest general findings that I pointed out in this paper was how relatively mild it could be. It raises a question about whether it is important to diagnose mild cases. Perhaps it isn't important unless they are a significant risk to others. Some children would have three or four coughing attacks, mainly in the night, perhaps for three or four weeks. Children quite commonly choke on a bit of food or drink that 'goes down the wrong way' and parents realise that they get over it quickly and no harm is done. It is a natural way the body gets rid of stuff in the windpipe that is not meant to be there. So when it happens repeatedly but the child is perfectly well they may think something is not quite right but it rings no alarm bells. And neither should it. If they knew it was mild whooping cough and there was a risk it might spread to a few weeks old baby and cause serious illness, then alarm bells might well ring but how many parents know that kind of thing? Very few. Would a GP given those facts see alarm bells ringing and arrange tests and antibiotics and school exclusion and worry everyone? Possibly not. Maybe if there was a young baby in the house the thinking might be different, but if not, what are the chances of it being passed on to a baby at the school gate or in the supermarket check-out queue. Very small, so perhaps not worth making a fuss over. These are the realities of life which mean that the bottom line is that the only practical form of protection is for everyone to be immunised, even if it is not very effective for individuals. But in the case of pertussis it creates all important herd immunity which keeps circulation down.

As a general observation on severity I would say that of the ones I diagnosed with whooping cough, about 20% had it mildly enough for it hardly to be noticed. They could continue at school for example without any teacher becoming concerned about occasional severe bouts of coughing. Then there were about 10% with very severe bouts of coughing that ended in being unable to breathe for 20 or 30 seconds. They frightened themselves and anyone who saw them, sometimes to the point of an ambulance being called. The remaining 70% had a very unpleasant time and it was obvious to everyone they had some kind of respiratory disease.

The diagnosis cannot be made early in the illness. It can take up to 14 days for the cough to become paroxysmal and this is a time when people are most infectious. It is a rather hopeless task to attempt avoidance of this disease using ordinary common sense measures.

A slight surprise was that adults had the same symptoms as children. Somehow we had been led to believe adults did not get it, so by implication they must have some inbuilt protection that would cause different symptoms if they did. But no, it is just the same. Not surprising given that this is the case with most infectious diseases.

More females than males caught it, about 10% difference in the young. In adults twice as many mothers as fathers caught it. This may be related to the amount of contact mothers have with their children compared with fathers.

The general distress caused by whooping cough was probably greatest in the loss of sleep experienced by parents, who were constantly up several times in the night to comfort and support a child having a coughing attack. No parent can easily listen to such an attack and not feel the need to go to that child. Indeed, it is right and proper that that should happen, not only for the psychological benefit of the child with a parent near but in case there is a risk of vomiting and inhalation, which could have serious consequences and can be prevented by holding the child prone.

To this day nobody knows what it is that makes whooping cough different from other coughs but different it certainly is. When anybody has heard it a few times it is easily recognised again. It is different because it is like a choking attack. It is as if the lungs want to empty themselves of air completely. So the cough goes on and on without taking a breath in until the lungs seem empty. Then there is a big gasp as air is sucked in and the process repeats itself several times, causing congestion of the face and even blueness if the gasp is a long time coming. Then it is over with perhaps a whooping noise and preceded or followed by vomiting and salivation. It is frightening to watch and terrifying to experience. One patient said it was like he had forgotten how to breathe.

Another strange thing that happens with whooping cough is that in the convalescent phase, which is from about four weeks from the start, just when it seems that recovery is slowly taking place, if you then catch a viral upper respiratory infection such as a cold virus, all the whooping cough symptoms, as they were at their worst, come back again. The poor sufferer thinks they have caught it again and that they are going to go through many more weeks of hellish coughing. This temporary relapse soon improves again, just as a common cold would, but it happens so frequently to people that it ought to be better known.

I cannot distinguish a paroxysm of whooping cough from an episode of choking after inhalation of a bit of food or drink. If I saw somebody in the street with such a thing I would assume it was choking, not whooping cough. But choking on a foreign body only happens once, and paroxysms of whooping cough happen several times a day, so there should be no real confusion.

I have already talked about common viral infections coexisting with the start of whooping cough and causing confusion, but secondary infections are common too. Sputum production is not a characteristic of whooping cough except for small amounts of sticky viscid mucus that is often coughed up, seemingly with considerable difficulty. If other bacteria get a hold in addition to *Bordetella pertussis*, they will induce the body defences that produce the yellow or green sputum that we sometimes associate with bronchitis, and that antibiotics are often prescribed for. The two conditions existing together can make diagnosis more difficult, as the pure, occasional paroxysm pattern may be masked by a frequent productive cough. In my experience this happens only in a minority of cases.

It took me many years to realise there was a characteristic to this disease which I had never seen properly described in any textbook or papers or reports but was actually a hallmark as distinctive as the cough itself. The closest one gets in textbook descriptions is, 'The patient can appear comparatively well between attacks'. As far as the patients I saw were concerned, and they probably represent what happens in most developed communities, they were *completely* well between attacks and you would never guess they had anything the matter.

Putting it another way, the main characteristic of whooping cough is that except for the occasional paroxysm, you do *not* cough!

I have developed a description of whooping cough that patients can use to diagnose themselves. It is:

> *Whooping cough causes bouts of coughing that usually go on for at least 3 weeks. In each attack there is a feeling of choking and suffocating. It frightens onlookers as much as the victim. These choking attacks occur on average a dozen times a day and between the attacks there is usually no coughing at all.*

There is nothing certain in medicine and what I have stated has to be interpreted in the light of individual circumstances. There are always other diseases that can seem very similar, so it is never a question of 'Oh, that sounds like whooping cough, off you go'. It is more likely along the lines of, 'It sounds like whooping cough but at your age we had better get an x-ray'. My 744 cases could have been 746, but one youngish man with symptoms consistent with whooping cough turned out to have a rare form of lung cancer and a middle-aged lady also with symptoms consistent with whooping cough turned out to have *Pneumocystis carinii*, an HIV-related lung infection. She did not know she had been infected by HIV until then.

*****

References

1. Hodgkin K. Towards earlier diagnosis: a guide to general practice: Churchill Livingstone; 1974.
2. Fry J. Profiles of disease. A study in the natural history of common diseases: E&S Livingstone Ltd; 1966.
3. Jenkinson D. Natural course of 500 consecutive cases of whooping cough: a general practice population study. BMJ. 1995;310:299.

Chapter 13
1990s. Whooping Cough Fades Away But New Diagnostic Tests Emerge

The NHS was given a big shake up in 1991 when Kenneth Clarke the then Secretary of State for Health introduced 'GP Fundholding'. He created an internal market by pricing hospital procedures and giving GPs a budget to purchase the services their patients needed. It was thought that such a process would benefit patients by reducing inefficiencies and incentivising innovation. In an organisation as large as the NHS it was inevitable, and we all knew it, that there was scope for improvement in many areas. Practices had to volunteer to join the scheme, and about half did so, us included.

We foresaw two benefits from joining. It would give us scope to provide new services for our patients and we were allowed to re-invest any profit that might be made into improving our premises. Not a single penny was allowed to go into our own pockets.

It was politically controversial, seen by some as helping productivity by introducing a profit motive to overcome inertia, while others saw it as undermining the ethos of the NHS which was free and equal health care for all at the point of need.

We four partners were split on the issue. Erl Annesley was wildly enthusiastic and couldn't wait to get started, Clive Ledger and I thought it was a good idea and would fully go along with it, while Andrew Wood was firmly against it. Andrew was an active member of the Liberal Democrat party and upheld his party's position on the matter. Andrew stood as the Liberal Democrat parliamentary candidate for Rushcliffe, which was our constituency, in the 1992 general election, against the aforementioned Kenneth Clarke, who was the longstanding MP for Rushcliffe. Nobody expected Andrew to win but he got a very respectable 20% of the vote.

BBC television came and filmed us Keyworth doctors discussing the Fundholding issue for 'Panorama' and gave us our '15 minutes of fame'.

The average NHS employee was, and to a large extent still is, oblivious of the cost of anything. Not necessarily because of lack of concern but because nothing is ever priced at the point of care, so nobody knows. Fundholding forced hospitals to put a price on the services GPs required, operations, outpatient attendances, laboratory

D. Jenkinson, *Outbreak in the Village*, Springer Biographies, https://doi.org/10.1007/978-3-030-45485-2_13

services; the list went on, but nobody could do it accurately because there were so many unquantifiable elements. The result was a large variation between providers which created a competitive market for GPs to exploit while hospitals got their pricing act together.

The new Labour government of 1997 abolished Fundholding, much to the relief of most of the NHS, as it distracted from the main business of the organisation and had produced no new money but simply redistributed it in a way that was not felt to be fair. But many changes made at that time were retained and built on with benefit.

We did quite a lot in Keyworth under Fundholding that was directly of benefit to our patients. We used a local nursing home for the intermediate care of some elderly patients. We held an orthopaedic clinic, an ophthalmology clinic, and employed a physiotherapist and a counsellor on our premises, and a nurse to visit the elderly at home. We bought a portable ultrasound scanner and built an extension to the Health Centre with two badly needed new consulting rooms with the money we saved.

Our day to day work, full of the mundane, the challenging, and the surprising, continued on just the same, with or without Fundholding, even with or without whooping cough, which was about to see some interesting developments.

Figure 13.1 shows the number of cases notified and vaccine uptake. The former drop as the latter increases. Nobody could look at it and doubt that immunisation was effective. It was all very satisfying and I had managed to get some useful research done. After the natural history paper I was not expecting to be able to write any more papers on whooping cough, but I knew I would continue recording any cases I discovered because that was my habit, and nobody knew what might turn up unexpectedly.

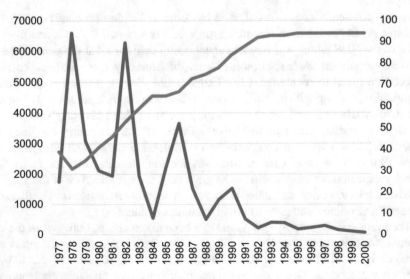

Fig. 13.1 Whooping cough notifications in England and Wales 1977–2000 (blue) and pertussis vaccine uptake per cent (brown)

The years after the fourth outbreak that ended about 1991 saw some important changes on the whooping cough scene that made diagnosis and investigation of the disease considerably easier.

Although by the mid-1990s it was accepted that whooping cough vaccine only very rarely caused severe reactions and probably never caused permanent brain damage, it was a frequent cause of minor reactions which were believed to be a consequence of it being a crude vaccine composed of whole bacteria with remnants of an endotoxin still present. The search had been on for a while to make a purer vaccine and in the mid-1990s they were being tested, and found, as expected, to cause fewer reactions. The purified vaccines contained up to five components, each one believed to stimulate the production of antibodies that played a part in developing resistance to the disease. Different manufacturers used different combinations of the components and they seemed to be equally effective when trialled and compared reasonably favourably with whole cell vaccines, so they were brought onto the market. These acellular vaccines had replaced whole cell vaccines in the USA by 1997.[1]

We took longer in the UK to start using acellular vaccine. Fears of another backlash against pertussis vaccine may have played a part. The vaccine refusal that had been so massive in the UK had been a much lesser issue in the USA, so acceptance there was easier. In the UK acellular vaccine was first introduced as an addition to the pre-school booster in 2001, which had previously only contained diphtheria and tetanus vaccine. It was a response to the realisation that pertussis immunity wore off more rapidly than had been thought, and it was therefore aiming to reduce deaths from pertussis in babies who were thought to catch it from older siblings.

Acellular vaccine was introduced for primary immunisation in the UK in 2004 because of the need to change from live oral polio vaccine to the safer inactivated version, which was also given in three injectable doses, so could be combined with diphtheria, tetanus and pertussis. The only vaccine available with that combination used acellular pertussis. From 2004, every child immunised in the UK received the acellular version of pertussis vaccine for the full course of four.

Improvements were happening on the diagnostic front too. Detecting *Bordetella pertussis* infection had depended entirely on obtaining live bacteria from the patient and culturing them on an agar plate where the resulting colonies could be recognised for what they were. It was uncomfortable to have done and unreliable. In the hands of the average doctor or nurse and laboratory it too often failed to give a positive result when the diagnosis was certain. There was now a new antibody technique available that approached the confirmation of pertussis infection from a different direction.

When our bodies fight infection they produce antibodies, complex protein molecules with a specific shape that lock onto the invader and mark it for destruction. In the case of *Bordetella pertussis* there are several different antibodies produced. An

---

[1] Acellular pertussis vaccine was designated aP, and whole cell pertussis wP to distinguish them.

ELISA[2] test can accurately measure the amount of such antibodies in a laboratory. A subject who has never been exposed to *Bordetella pertussis* or the vaccine will have no appreciable antibodies to its various components. After exposure and invasion, antibody levels rise within a week or two and will remain high for months then slowly decline. Importantly, the body remembers how to produce these antibodies so that if there is another attempt to invade, they are produced rapidly enough to repel the invader without the person being aware, but the evidence will be in the blood as raised levels of antibodies again. When this investigation was first introduced into pertussis antibody analysis, knowledge of how to interpret the levels of the different antibodies was incomplete but it quickly improved.

After my traumatic viva in 1996, Professor Elizabeth Miller told me she was looking for interested GPs to provide blood samples from pertussis patients in order to develop a blood test for it. I was delighted to participate naturally, and so I came to benefit from the very early clinical application of these blood tests which became increasingly refined.

Elizabeth Miller and Douglas Fleming and others used that new testing technique in 1996 and 1997 to investigate *Bordetella pertussis* antibody levels in possible pertussis patients in a Birmingham practice with a population about the same size as Keyworth's. I was fascinated by their results that suggested an infection rate 80 times greater than the prevailing notification rate [1], although that actual notification rate had fallen to a very low level in England and Wales (but not in Keyworth). Caused, in my view, by loss of expertise in diagnosing it.

A similar piece of serological research had been done in the USA in 1995 [2]. The investigators analysed the blood of 51 adults over five years. They were looking at antibodies to *Bordetella pertussis* and measuring the levels over the period. What they found was startling. The majority had had a rise in antibody levels and very many had had two or more separate rises. None had knowingly had whooping cough. Although it was indirect evidence, it suggested that *Bordetella pertussis* was infecting people regularly and boosting immunity without being noticed. Not only that, it also implied that the immunity acquired from becoming infected was not very long lasting. It had previously been thought that immunity from natural infection was highly durable, possibly lifelong.

Increasing understanding of the various pertussis antibody changes and improved investigative techniques led to a simple blood test being made available that was reliable enough to be used as a routine NHS test.[3] It was introduced in the UK in 2002. The test can be done on a single sample of blood taken after two weeks into the illness and it will tell with fair precision whether you have had a recent *Bordetella pertussis* infection or not, and it can be done and give a valid result for several months after the initial infection. The same test can now also be done on oral fluid,

---

[2] ELISA is an abbreviation for enzyme-linked immunosorbent assay. Now reduced to a three letter acronym EIA.

[3] The simplified test, which detects 90% of cases, measures pertussis toxin immunoglobulin G in a single sample.

although the false negative rate is somewhat greater. The latter test is usually reserved for children.

Another breakthrough came with advances in DNA technology. By the mid-1990s PCR[4] tests were revolutionising many fields of science, the best known being DNA fingerprinting in forensic investigations. The test can identify fragments of DNA that can be seen to belong to a bacterium such as *Bordetella pertussis*. The advantage is that only a trace is needed and may be found even after they have all been killed or in numbers too low to detect by culture. This test increased the chance of confirmation of the diagnosis markedly but was expensive when introduced in 2002. The cost of such testing has plummeted and is now available as a test in general practice. It can be done on a dry throat swab, which is a routine easy test, instead of an uncomfortable per-nasal swab, although the latter or nasal aspirate is best. This same technology is also used to investigate how *Bordetella pertussis* changes over time, possibly in response to vaccine-induced selection.

*****

There were almost 3000 notifications of whooping cough in 1997 in England and Wales, and on the graph it shows as a slight bump (Fig. 13.1). This was the highest number there was going to be in the UK for 15 years until 2012. How did the experts interpret this?

It was seen as proof of the effectiveness of the immunisation programme. There was no reason to think otherwise. We seemed to be enjoying pertussis almost flatlining, but puzzlingly, one of the main objectives of the programme, cutting infant deaths, was little changed. There was still a handful of deaths every year and for every death there were roughly a hundred times as many infants who recovered thanks to paediatric intensive care. On the face of it this had to mean that the infection was still out and about, but it was not showing up as notified cases, so where was it?

These issues were occupying some of my thinking round about 1995 and I was not expecting to see much whooping cough at all, but as the years went by and I was still diagnosing it with laboratory confirmation fairly frequently, I became increasingly puzzled and frustrated because *I* kept seeing it, but others seemed not to. I knew my pick up rate was higher than average, but I had made a special study of the disease, so of course I was going to detect more. But when I compared my diagnostic rate with that of the whole country, I realised there must be thousands of people out there who were getting whooping cough but were not being diagnosed. I knew this because in 1995 and 1996 Keyworth was notifying 100 times more cases per unit of population than the rest of England and Wales. So in 2000 I turned my hand to what I thought might be a logical remedy for this diagnostic paralysis. I would create a website to help patients diagnose themselves, as I knew their doctors weren't going to do it and it must have been a big problem for them.

*****

---

[4] Polymerase chain reaction. A trace of DNA is copied over and over again until there is enough to measure.

References

1. Miller E, Fleming DM, Ashworth LA, Mabbett DA, Vurdien JE, Elliott TS. Serological evidence of pertussis in patients presenting with cough in general practice in Birmingham. Commun Dis Public Health. 2000;3(2):132–4.
2. Deville JG, Cherry JD, Christenson PD, Pineda E, Leach CT, Kuhls TL, et al. (1995). Frequency of unrecognized Bordetella pertussis infections in adults. Clin Infect Dis. 1995;21(3):639–42.

Chapter 14
2000. Whoopingcough.net. Overcoming Diagnostic Paralysis

The internet was invented in 1989 and boomed. In 1996 I had come across a website in the USA that called itself a virtual reality hospital. It invited experts to post information to help patients. The internet was new and primitive compared to what we are used to now. I put some text up about whooping cough and was surprised, not just to get responses, but responses from patients who were obviously suffering from whooping cough but were not being diagnosed. They seemed to be being subjected to expensive and unnecessary investigations because whooping cough was not being considered, or even being denied as even possible by their doctors. This confirmed to me there was a need for help for such people.

I had created a website for the Keyworth practice the previous year, 1999, so I was a little way up the learning curve. www.whoopingcough.net was a simple affair with a few information pages and an email inquiry page at the start. It went live in 2000 (Fig. 14.1). My aim was to provide comprehensive information about whooping cough as I saw it, and to help sufferers get it diagnosed, if only in the English-speaking world for those with internet access. It was a key development in my involvement with the disease as I came to realise from the feedback just how valuable people found it and how influential it was in changing perceptions. Perhaps more importantly it confirmed to me that there were large numbers of people out there with it going unrecognised, as I had suspected, but also that it was not just in the UK but in Australia, New Zealand, the USA and Canada, where they appeared to have the same problems with occurrence in adults and difficulty with diagnosis.

A great source of satisfaction for me was diagnosing adults with whooping cough and seeing the effect the diagnosis had on them. It was not greeted with shock and despair. Surprise, certainly! The almost universal reaction was relief and profuse thanks, because they believed such a terrible cough was most likely some serious or fatal lung disease. Every year there were thousands of adults in the UK alone, seriously worried about a suffocating cough that lasted months with no answer to it. Of course, almost all would see a doctor. They would be examined, often have blood tests and a chest X-ray too, and these tests would all be normal. The doctor would

© The Editor(s) (if applicable) and The Author(s), under exclusive license to
Springer Nature Switzerland AG 2020
D. Jenkinson, *Outbreak in the Village*, Springer Biographies,
https://doi.org/10.1007/978-3-030-45485-2_14

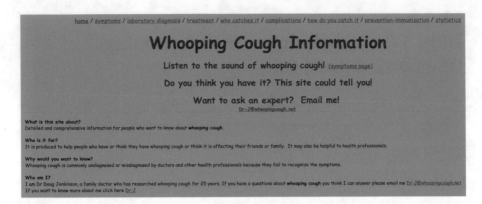

Fig. 14.1 The landing page of whoopingcough.net when it first appeared in June 2000

reassure them that it was nothing serious and it was probably just a virus and they all recovered in due course. It was frustrating because I knew a diagnosis would save them months of anxiety.

Full recovery from whooping cough being the rule in older children and adults is the main reason it is so difficult to raise the profile of this disease.

I had three areas of concern about the failure to diagnose whooping cough. Firstly, undiagnosed whooping cough is a serious risk to infants under four months old who might come into contact with this clear hazard. Secondly, I felt it wrong that a lot of adults were suffering anxiety and lack of diagnosis unnecessarily. Thirdly I was embarrassed that doctors were failing to diagnose an easily recognisable disease; it was not going to help the profession's reputation.

Over the years I had written articles in medical newspapers, sent letters to journals and talked to groups of doctors in an effort to improve the rate of diagnosis, but with apparently little effect. The main problem before the millennium was that there was no confirmation to be had from a suspected diagnosis, so doctors received no positive feedback by confirmation, which is so important for diagnostic confidence. Shortly after the millennium it became possible to confirm the diagnosis in most cases with a blood test, but its availability was not publicised and so not known about by the majority of GPs until several years later.

The site was aimed at the people I felt most sorry for, the undiagnosed sufferers. The heart of it was the sound files I put there. It seemed an almost miraculous facility to be able to click on a page and hear the sound of different examples of children with whooping cough, some whooping and some not. Soon after those, I put on the sound of one of my adult patients coughing and vomiting. It was nauseating just listening to it! For many years these sound files were the only way of hearing whooping cough. The same ones are still there now.

The recordings were all of my own patients. Some I had asked the parents to record for me as samples, but most I had obtained for the purpose of diagnosis. I would lend the parents a mini tape recorder and ask them to record the cough, because the unwritten law of whooping cough is that coughing will never occur during a consultation!

Sound files were rare on the internet at that time and it was very much slower than it is today. Still images would download reasonably quickly but even a short sound clip could take 20 seconds. Internet browsers were simpler too, and could not handle the complexity we are now used to, but simplicity was just fine for my purposes.

I constructed several pages covering symptoms, diagnosis, treatment, prevention etc., the information I thought people might require to diagnose themselves or others. I was soon able to make a page of frequently asked questions too, because there was an email comment page that attracted hundreds of comments. Feedback was vital if I was to understand what was going on in the wider world of whooping cough.

I was fortunate to be able to design the website myself. It was not particularly difficult as by then Microsoft had devised a friendly web design package called 'Frontpage' which I could use.

Contrary to what people might expect, it is not expensive to set up a website. A few pounds will buy you a domain name, and then you pay a small registration fee every couple of years. The most expensive bit is the service provider onto whose computers you load your website. A hundred pounds or so a year for a simple one. I was quite happy to pay this for the satisfaction it gave me.

I had to be careful to follow the UK General Medical Council (*GMC*) guidelines on doctors and the internet. They are always ultra conservative and crossing them can have very serious consequences for a doctor, so following their strictures was vital. They considered it impossible to conduct a medical consultation over the internet so stressed not setting up anything that looked like there was a professional doctor and patient relationship. I took great care about this, putting a specific statement on the advice page about it. I hoped they would not look unfavourably on such an altruistic site.

Over the years I have had thousands of email comments confirming that the description I gave of whooping cough enabled patients to diagnose themselves. They then went to their doctor with their diagnosis suggestion, sometimes also with my 'printout for doctors' with them. In the early days of the site and even up to about 2014, the common outcome, particularly in the USA, was a refusal of the doctor to accept the diagnosis. This has gradually changed. Since about 2015 I have been receiving fewer email comments complaining of not being listened to. This is a big turnaround and it is gratifying to think I may have helped change perceptions.

Many GPs get a sinking feeling when patients walk in with a handful of printouts from the internet. Knowing I have contributed to this does not seriously worry me. I have never felt entirely comfortable with the traditional doctor/patient hierarchy, where the doctor talks and the patient listens, although it may have been appropriate historically and in some settings today. I think there should be a partnership relationship between doctor and patient and if the patient sets the agenda with a printout, I see no reason why that should not be the starting point of the interaction. I can also see perfectly well that it does not sit easily with the frequently time-pressured reality of many working medical environments.

One of the things I discovered from the website emails that I had not appreciated was the effect whooping cough can have on the voice. It seems to be mainly singers who complain about it. It changes a good singer to a useless singer for a variable length of time and commonly lasts for months. It can be a year, or sadly I have heard from people for whom it was permanent. It can have serious consequences. One of my correspondents was a professional international choirmaster who had to cancel tours. Voice change is quite common and has been recognised for a long time. I came across a letter in a professional journal written by an ENT specialist in 1932 describing voice changes with pertussis and also pointing out that pertussis was not uncommon in adults (Fig. 14.2).

I have noticed that general medical websites, of which there are now many, have gradually changed to reflect the kind of information I have been publicising. They quite rightly put their emphasis on the danger to babies and the importance of immunisation but also now accept that it affects teens and adults too, often with encouragement to get booster shots (which are not available on the NHS except in pregnancy, and for certain healthcare workers). These websites now tend to reflect the same information I have had on whoopingcough.net since 2000. One of them copied my text almost word for word. That was just fine; I was flattered. The key points are now included, which is the most important thing. They like the concept of the '100-day cough', which is a great hook, and I think I was the first to publicise it on the internet. It is supposedly what it is called in China.[1] But most important of all is the 'long periods with no coughing at all', which to my mind is the most specific symptom of whooping cough and is gradually becoming better known.

An important piece of advice that I emphasise on the site that others are not including yet is to video a paroxysm. This sometimes means recruiting somebody else to do it, but the purpose is to show it to your doctor, who will presumably not then fail to consider whooping cough as a diagnosis. It overcomes the single biggest barrier to diagnosis, lack of evidence. With a smartphone the facility to do this is available to almost everyone. The phone can be put down with the video running during a paroxysm; it is the sound that is most important.

There is now good whooping cough information on many websites, and YouTube has lots of videos of it, including some of my correspondents sent me and permitted me to use. I applaud their public spiritedness. They all said, 'By all means use it if it helps people to understand this awful disease'.

The other sites on the subject are now numerous, simpler, and therefore easier to digest. Whoopingcough.net is detailed and comprehensive, so can be rather overwhelming for the modern internet user with a short attention span, but very useful for people who are searching for in-depth information. There is quite a lot of work involved in keeping a modern look to it and keeping it up to date. I have given it a

---

[1] Early Chinese writings about a '100-day cough' are almost certainly referring to a very different disease with a high mortality in adults.

CLINICAL RECORDS

WHOOPING-COUGH AND LARYNGITIS.

By DAN McKENZIE, London.

HAVING to-day seen my third case of whooping-cough in adults in the last three weeks, I am impelled to send this little note to draw your readers' attention to a common source of mistaken diagnosis. The mistake is all the more likely to occur as the period prior to the formation of the characteristic cough is, I find, frequently marked by an intractable acute, or subacute, laryngitis, the nature of which is apt to be perplexing, naturally enough.

These patients are frequently sent to the laryngologist as they may not have been in contact with children suffering from the disease, and the true diagnosis does not at once leap to the eye of the ordinary practitioner, nor yet we must admit, to the eye of the laryngologist either, as the history of the illness and the patient's description of his cough are apt to lead our wits astray in a search for some out-of-the-way cause of "glottic spasm."

But the suspicion of whooping-cough, once raised, soon becomes a certainty, and our diagnostic skill is vindicated.

A former attack in childhood, by the way, does not exclude the possibility of the disease recurring in adult life.

I finish with the remark that the treatment of whooping-cough has changed somewhat since I was in general practice thirty years ago, but the results are the same. The cough still lasts for seven or eight weeks, and furnishes a most objectionable experience.

Fig. 14.2 Letter about voice change in whooping cough and occurrence in adults from Dr. Dan McKenzie, August 1932, in the *Journal of Laryngology and Otology*

facelift recently using WordPress and added automated translation into most languages. This has also enabled me to add myself to the blogging community as another way of spreading information about whooping cough.

I have never knowingly had whooping cough myself so cannot describe it from personal experience, and I have already tried to give indications of what it is like, but inadequately, compared with stories below that I have been sent by people. They were sent as comments to help other people and I think their voices are compelling, informative and highly descriptive of the typical problems encountered. The following are choice examples.

I am still suffering from the effects of suspected Whooping cough—five or six weeks in now. I struggled to remain at work having had little sleep due to the early hours coughing, retching, then constant vomiting. At this stage I had visited the GP twice; one Dr stated I had a common cold/upper respiratory virus, the other concurred but thought the symptoms matched Whooping cough—if I was 10 months old, not 47 years! The scariest thing was after coughing and vomiting I could not draw breath—not just for a couple of seconds, much longer—it was like someone had placed clingfilm over my face. My wife is a nurse and she was very concerned. After a third consecutive night of this level of suffocation, my wife insisted on driving me to the ED of my local hospital at 05:00 in the morning. Chest X-Rays were clear, blood oxygen normal. The ED Dr was very sympathetic and knew I was 'not right' but could only say the same as the GP—upper respiratory virus. He did refer me to the ENT registrar, who examined me with a nasendoscope and found some adenoid inflammation. He prescribed an antacid to prevent reflux burning my airway. I was forced to take a week and a half off work and visited the GP twice more; firstly to be prescribed Amoxycillin—then the final time after I returned to work (not that I was feeling much better, but sick leave is something I have rarely used) my GP decide to do blood tests and send them to London for Whooping cough analysis. Ironically she told me I may have been past the stage where it could be identified anyway! Enduring an awful time back at work and probably should have taken more time off. I have never experienced anything like this before and can sympathise with other comments regarding the viscous nature of the mucus and the alarming nature of the breathlessness following severe coughing bouts.

Thank you for your website. It has been invaluable and kept me sane.

That was the worst part of the disease; everyone thought I was making it up (except my girlfriend that had to endure me waking up in the middle of the night unable to breathe).

I had pertussis last February, and like many people who've contracted it, I didn't know what it was. Despite having consulted five doctors (two GPs, two ER physicians, and one lung specialist), I was left in the dark. Even lab results came back negative for anything other than the common cough. It wasn't until I stumbled upon your website when I realized what I was dealing with. The symptoms and audio files found on your website matched my condition to a tee!

I should also say that the term "whooping cough" is a grave misnomer. This disease should really be called "gasping cough" since that's what really happens. You're literally gasping for air because it feels like there's suddenly not enough of it around you. You're drowning and you're not even in water!

It is now in its 87th day with no silver lining in sight! What no one really appreciates is the long tedious nights, anxiously coughing, on & on, until the dawn! Cough medicines are totally useless.

I then did some of my own research and came across your site after googling my symptoms and coming up with possible whooping cough—I took your advice and

got my husband to video an episode, lasting ten minutes—a second GP watched
8 seconds of it and agreed that it looked like whooping cough.

11 days ago after passing out in the early hours I went to see my, thankfully
infrequently visited, GP of many decades. In great detail I described the nature of
the 'cough like no other'. Every time I have seen a doctor, I have described it so at
length (reinforced by my wife) that within a few seconds of being OK(ish) about 12
times a day, I would succumb to the violent coughing like no other & often the
TOTAL blockage of my airways, unable to force air in or out of my lungs, the very
strange noises & light headedness that accomplished each attack.

I spent the rest of the night self-diagnosing on the internet. When I played the
Male with whooping cough making loud whooping sound track my wife rushed in
convinced I was having another attack. As ill as I was we were both relieved to
KNOW what I was suffering from.

Initially it was a very 'chesty' cough for a number of days however for the last
2 weeks my symptoms are EXACTLY as you describe. I can have long periods of not
coughing but when I do it can go on for ages. I have a very distinct inspiratory stri-
dor just like the audio on this site, often leading to vomiting, I feel dizzy after each
episode, and often cannot talk properly following an episode of coughing and have
not slept properly in weeks (lying flat exacerbates coughing). I have lost 4 kg in this
time because swallowing is sometimes difficult and vomiting always likely.

The part that we are really struggling with is the paroxysms. He has been having
them almost every hour and his lips turn blue. He just looks up at us with these terrified eyes when its starts and then coughs, chokes, gags until he eventually sneezes
or just stops and goes completely limp in our arms so exhausted from what he just
had to go through.

I am a teacher and currently display all symptoms of whooping cough including
feeling I'm suffocating at least twice an hour. I took myself to A&E yesterday and
was told my chest was clear and I was fine.

After a night of terrifying my family I rung the out of hours to who listened to my
symptoms and asked me to go to the clinic. I was told I was fine, they could only
diagnose whooping cough with a tube down my throat.

I explained I was in contact directly with 800 children per day and was told if I
wanted time off I'd need to self-sign and that the pupils had all been vaccinated and
were surrounded by viruses every day.

I left in tears…I never take time off work…I never usually go to a doctor.

Please, please warn people who read your website that their GP is not likely to
help them.

Some of these emails are clearly written at a time of great stress and may not be completely objective in their accounts or interpretation of the advice they were given, but I have included them to show what it is like to have this disease and have it dismissed or undiagnosed, albeit with the best of intentions.

I have no way of quantifying the impact of whoopingcough.net and whether it played a part in bringing the disease back into the general medical consciousness. Its clear aim was to help sufferers diagnose themselves, tell their doctor and, if possible, get tested.

Most site users were from English speaking countries, principally the USA, Canada and Australia until quite recently. When I look at monthly statistics now, I find that almost every country in the world has been represented as a result of the pages now being machine translated into most languages. Currently, if a user's browser is in Japanese for example, the Japanese translation will be selected. This facility, which is now widely used, could change many things.

In the early years when it was just about the only site on the subject it was getting one to two thousand visitors a day. In October 2015 there were over 1500 per day. In the whole of 2016 there was a total of 423,000 visitors. There are now many excellent sites sharing whooping cough traffic, so numbers are now smaller, but there is another reason too.

A big drop in traffic happened in mid-2018 when Google changed its ranking algorithm for sites that give health or well-being advice, or took money. It calls these 'Your money or your life' sites. To rank highly, such sites have to demonstrate, to the satisfaction of Google's (undisclosed) criteria, high levels of expertise, authoritativeness and trustworthiness. It is a logical, reasonable and laudable policy, I have to concede, but whoopingcough.net cannot attain high ranking because it is not affiliated to any institution of repute. Google sees it as a one-person site that could be eccentric because there is just one person in charge, so I have dropped from the first page to the third or fourth page. This makes an enormous difference to the number of 'clicks' you get. Search engines wield enormous power in this respect.

I know the site still provides a useful service as the emails still come in saying it is the only really comprehensive site on the net on whooping cough, and visitor numbers are building up again as more backlinks[2] are created. So I intend to keep it going for as long as I can.

---

[2]A backlink is a reference in another website or blog to your site. Lots of backlinks are good evidence of quality and improve your Google ranking.

Chapter 15
The Early Noughties. The New Tests Aid the Re-discovery of Whooping Cough

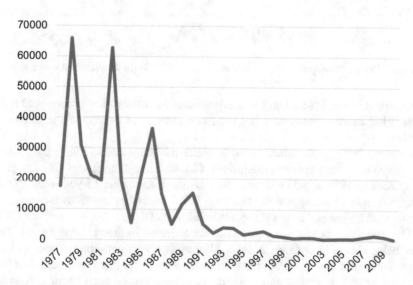

Fig. 15.1 Notifications of whooping cough in England and Wales 1977–2010

After the millennium I could see clearly that the medical profession and the general public had been, and still were, oblivious to the fact that whooping cough was quite common and almost invariably missed as a diagnosis. Figure 15.1 shows the flat-lined notifications after the millennium. But because it was not life threatening to the patients I saw, and the national death rate in infants was low, it was a situation I could live with. I thought that perhaps I knew a little better, but it was not of immense importance. I also knew from whoopingcough.net that under-recognition was a worldwide problem and not just happening in my small corner of the world. I was content to believe that the truth would eventually become clearer to everyone, but

© The Editor(s) (if applicable) and The Author(s), under exclusive license to
Springer Nature Switzerland AG 2020
D. Jenkinson, *Outbreak in the Village*, Springer Biographies,
https://doi.org/10.1007/978-3-030-45485-2_15

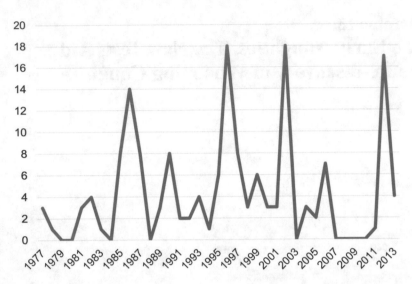

Fig. 15.2 Cases of whooping cough in Keyworth 1977–2013 in the 15 years and over age group

nevertheless it was frustrating. I wanted to make the information known but I didn't know what more I could do other than my website to overcome the diagnostic paralysis.

I was particularly concerned about the pattern in teens and adults, as they now, in my experience, were the ones catching it. I looked at my figures for this age group going back to 1977 and they showed that the number of over 15-year-olds affected in Keyworth had been apparently low until 1986, when they suddenly increased and remained at that raised level subsequently (Fig. 15.2).

The frequency in this age group was not reflected in the national figures. They were being almost completely missed. Although older generations of doctors knew adults got it, and that you could have it more than once, this knowledge had been temporarily lost by mainstream doctors. The reasons such facts become forgotten are probably multitudinous, but I suspect it might sometimes be because we become carried away by our own immunisation hyperbole.

When a new vaccine is introduced it is important to persuade everyone eligible to have it because that way achieves maximum benefit. Before all the details of a new vaccine are known, which can only come after many years of experience with it, the tendency will be to exaggerate its benefits to achieve a high uptake. Thus, in the case of pertussis, the suggestion would be made or implied that the protection was very good and therefore the disease would be immunised out of the susceptible population. Those promoting it in that way would come to believe it too because there was no information to the contrary.

The newer generations of doctors may never have seen a case or been taught the difficulties of diagnosing it or understood that it was not just a disease of children. It is the fate of a declining disease to be given less prominence in an increasingly crowded medical curriculum. The doctors who understood about whooping cough had mainly retired or died and, if not, were probably not teaching medical students or training GPs. Whooping cough also suffered from the fact that it is one of the few diseases that can only be clinically diagnosed by sound, an aspect that cannot be adequately dealt with in textbooks.

I was not the only one diagnosing whooping cough. My medical partners in Keyworth certainly were, and several of my local medical colleagues would from time to time tell me they had seen cases when we met up at meetings. We would bemoan the fact that it so often got missed because it was not thought about or was considered to be just a childhood illness. The occasional letter had come my way too, from doctors who seemed as enthusiastic as myself, having realised I was onto something. These would usually follow a piece I had written in one of the weekly medical newspapers like 'GP', 'Pulse', or 'Doctor'.

Figure 15.3 illustrates the under-diagnosis of whooping cough in this period between 1995 and 2006. The incidence of infectious diseases has been traditionally expressed as the number of cases per 100,000 population in any given time period, usually a year for statistical purposes, but at the start of an epidemic it is more usual and useful to state the number of new cases per 100,000 population per week or month. We see this during a flu epidemic, and it tells us whether it is waxing or waning.

If we look at the annual notification figures for whooping cough and express them, not as a total, but as the number per 100,000 population, the shape of the graph is the same[1] but it becomes possible to compare different populations. To arrive at the figure, we divide the total notified by the number of 100,000s in the population of England and Wales (about 55 million), which is 550.

If I do the same transformation for the population of Keyworth (11,000), the rate per 100,000 can be compared with the rate for England and Wales on the same graph.

It can be seen that up to 1992 the graphs correspond in shape but the Keyworth line is roughly up to 10 times as high. My explanation for this is that in that era I must have been diagnosing twice as many in Keyworth and notifying them all, whereas in general only 20% are notified. After about 1992 the England and Wales line flattens out until 2011 as a result of the diagnostic paralysis.

If the differences between the two are expressed as a ratio on a bar chart, the result is even more dramatic as can be seen in Fig. 15.4.

The bar chart starkly expresses the extent to which I diagnosed more in Keyworth. From 1995 to 2006 the difference is enormous, when it appears that on average I diagnosed about 100 times more. The most extreme year is 2002 when the Keyworth practice notified 249 times as many cases of whooping cough per 100,000 population as England and Wales. There were 44 cases diagnosed and in 23 of them there was also laboratory confirmation. Keyworth Health Centre was responsible for five per cent of all the notifications of whooping cough in England and Wales in 2002!

I think this is powerful evidence that doctors in general had stopped diagnosing whooping cough in the years 1995 to 2006. Putting it bluntly, for the majority of

---

[1] As long as the populations remain numerically the same.

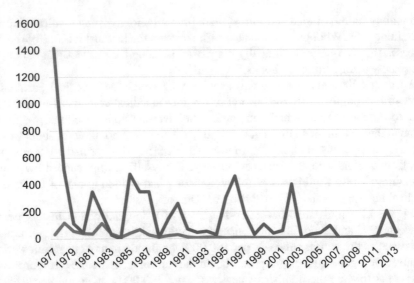

Fig. 15.3 Notifications of whooping cough in England and Wales (brown) and Keyworth (blue) 1977–2013 as cases per 100,000 population

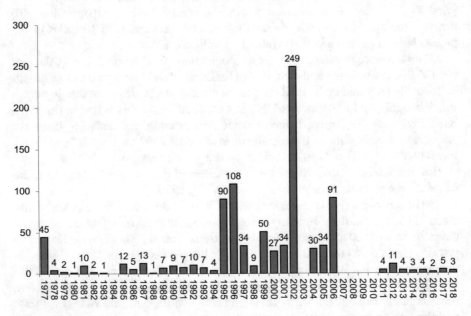

Fig. 15.4 Ratio of whooping cough notifications per unit population Keyworth v. England and Wales 1977–2018

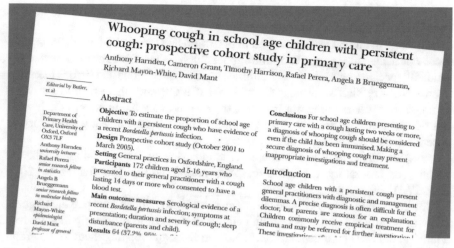

Fig. 15.5 Anthony Harnden's paper in the *BMJ* July 2006

doctors, whooping cough had ceased to exist, and there would be a 10- to 15-year gap in the national figures before whooping cough came to be recognised and diagnosed again. There were several factors that led to this rehabilitation.

I am sure whoopingcough.net played a part. Since starting up it had been accessed millions of times. Then in 2006 came the Harnden paper (Fig. 15.5).

Anthony Harnden and a team from Oxford [1] came along with a fascinating and revealing paper in 2006 that confirmed that there really was quite a lot of whooping cough in children who were generally perceived to be too old to catch it. They tested 5- to 16-year-olds who had a persistent cough and had attended one of 18 general practices in Oxfordshire between 2001 and 2005. They were able to find 172 with this complaint and who agreed to blood testing. A high proportion had whooping and vomiting. Their blood was tested for pertussis antibodies and 37% had evidence of recent infection.

This paper was widely read by doctors and achieved national newspaper coverage too. Harnden followed it a couple of years later with an educational piece in the *BMJ* about pertussis [2]. It was after the 2006 paper was published that doctors seemed more ready to accept the diagnosis in older children and adults, and do the now easily available blood test if it was suspected. It was also about this time that a doctor would find that if a notification was sent off about a case of whooping cough, a form was received back, requesting a blood sample along with other details of the patient history such as immunisation status. The Public Health Laboratory Service had introduced this 'Enhanced pertussis surveillance programme' in 1994 and the advent of the blood test in 2002 was a breakthrough that put pertussis surveillance on a firmer scientific footing.

The blood test confirmation method, and later the addition of oral fluid testing for 5- to 16-year-olds in 2013 and 2- to 16-year-olds in 2018, made all the difference to obtaining a diagnosis. At last, suspected cases could be confirmed. The main impact of such testing was that cases thought too mild to be pertussis would come back as positive and the lesson would be learnt.

Although the blood testing method of detection had been rolled out in 2002, GPs were not made fully aware of it until about 2006, when the existence of the test gradually became better known. I think it was so rarely requested before then that even laboratory staff were unaware of it, so if a GP rang them to find out what sample bottle was required, for instance, they would very likely be told the lab did not do that test. I personally recall some difficulties persuading my laboratory to accept samples for testing in those years.

Now that these tests had become the standard tests for whooping cough, the number of laboratory confirmed cases overtook notification figures in 2010 (Fig. 17.1).

The 2006 Harnden paper gave me an opportunity to publish my up to date figures showing continuing pertussis cycling activity in contrast to official figures. I wrote a letter to the *BMJ* commending Harnden's findings and relating them to the graph I included [3]. It was printed prominently, and it was my hope that it would get noticed, but I saw no sign of that (Fig. 15.6). At no point did I get a hint that my data were valuable, but I can understand that. If it had been possible to publish it as a paper it would have been better, but there was a large obstacle to that. Principally it was my reliance on clinical diagnosis, which was too subjective in an evidence

Whooping cough is quite common and can be diagnosed clinically

EDITOR—Harnden et al, with the help of a recently available blood test, have gone some way to confirm that whooping cough is still about.[1] Since 1977, I have, with the help of my practice colleagues, been recording every clinically diagnosable case of whooping cough (based on a minimum of three weeks of paroxysmal coughing). This year so far we have seen six cases in a practice of 11 000. In 2002 we recorded 44 cases. Ten had blood specimens tested; nine were positive and the 10th was lost. Twenty three had pernasal swabs taken; 14 were positive. This small practice was responsible for 5% of all whooping cough notifications in England and Wales in 2002. This surely cannot be because there is more whooping cough where I work.

I have had four papers published in

data are correct, is just as common as it was 25 years ago after the vaccine scare settled.

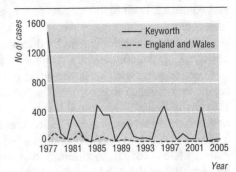

Incidence of whooping cough in Keyworth practice and in England and Wales both equated to 100 000 population

Fig. 15.6 Part of my *BMJ* letter contrasting Keyworth with England and Wales 12th August 2006. A similar graph is in Fig. 14.3

based culture, and secondly was the fact that it was just me, a one person outfit, with the possibility of there being unrecognised distortions or biases, or even worse. I knew, of course, that I was, and always had been, as honest and objective as I could in my data gathering and analysis.

A further boost in the raising of medical awareness came when Ros Levenson wrote an article in the *BMJ* as a layperson which was one of a series they were doing at the time under the umbrella title, 'A patient's journey', in her case a journey through whooping cough [4]. It was a thoughtful and clear account of the now familiar (to me) scenario of strongly suspecting one has whooping cough, but being unable to have it diagnosed, and even ending up seeing a chest physician and it still not being diagnosed! She described how whoopingcough.net had been instrumental in her self-diagnosis (Fig. 15.7).

Looking back, I feel sure that 2006–7 was when the tide turned, and awareness and recognition of whooping cough started to regain the lost ground as a result of diagnostic tests, new research and publicity. But there was something else happening that was going to seriously confuse the picture. The USA and Australia were reporting steadily increasing numbers of cases of pertussis in older children and teenagers and it was being blamed on the acellular vaccine.

PRACTICE

For the full versions of these articles see bmj.com

A PATIENT'S JOURNEY
Whooping cough

Ros Levenson

London

Correspondence to:
Ros Levenson
ros@roslevenson.demon.co.uk

BMJ 2007;334;532-3
doi: 10.1136/bmj.39120.556296.AE

Recently I had whooping cough. This has now been confirmed by the results of a blood test. I was in little doubt about the diagnosis from the beginning. However, my experience suggests that, despite considerable literature on the prevalence of whooping cough in adults, the diagnosis is still not one that doctors readily think of when they are faced with an adult rather than an infant or child.

I can well understand why the general practitioners I saw in the first couple of weeks did not diagnose whooping cough at once. I am aware that the duration of the symptoms, as well as the nature of the symptoms themselves, is important, and there were more obvious diagnoses that sprang to mind. However, when I experienced the first two or three terrifying paroxysmal coughing sessions, I knew it was a cough unlike anything I had had previously and said so. It also reminded me of my daughter's whooping cough when she was a toddler, many years ago. I was also very struck by the information on the website of Dr Doug Jenkinson (www.whoopingcough.net), which gave descriptions that matched my symptoms exactly. As a result, I

Tracheal epithelium showing *Bordetella pertussis* (green) lodged between the cilia that protect the respiratory tract from dust and particles. The micrograph also shows a region of epithelium in the foreground where the cilia have been flattened by the bacteria

sufficiently open to other, additional possibilities. Inter-

Fig. 15.7   Ros Levenson describes a patient experience of whooping cough in the *BMJ* March 2007

*****

Fig. 15.8 Keyworth Primary Care Centre opened by the Right Honourable Kenneth Clarke MP, April 2007. The parish church with its unique lantern tower is in the background

2007 had been a time for celebration for the GP practice in Keyworth. After many years we were moving out of our tiny and cramped prefabricated building into the magnificent new premises known as Keyworth Primary Care Centre that had been built on the old village car park. The old building was flattened and turned into a new village car park. It was a great boost for everyone, and the logistically difficult and rapid transfer from old to new over one weekend had been accomplished smoothly thanks to the organisational skill of our practice manager Michelle Broutta (Fig. 15.8).

We had been desperate for more accommodation as our increasing workload had required us to take on more GPs: Jill Langridge in 1995, Corinna Small in 2002 when Erl Annesley retired and Jim Hamilton in 2004. It was an opportunity to create new services. Jim Hamilton took on skin surgery in the practice and was later joined by Neil Shroff in 2005 who developed surgical services for skin cancer in primary care.

*****.

References

1. Harnden A, Grant C, Harrison T, Perera R, Brueggemann AB, Mayon-White R, Mant D. Whooping cough in school age children with persistent cough: prospective cohort study in primary care. BMJ. 2006;333:174–7.
2. Harnden A. Whooping cough. BMJ. 2009;338:b1772.
3. Jenkinson D. Whooping cough is quite common and can be diagnosed clinically. Letter. BMJ. 2006;333:352.
4. Levenson R. A patient's journey: whooping cough. BMJ. 2007;334:532–3.

Chapter 16
Late Noughties. 'Resurgence' of Whooping Cough in the USA and Australia Puts the UK on Alert

It probably seems strange to many people that nobody knows precisely how many cases of whooping cough there are on a worldwide or even national basis. But what is being counted? And who is counting accurately? Generally not the doctors or nurses who may diagnose it unless they get paid to do so, which is rarely the case and even then the reward is hardly ever worth the effort. Most countries have a system that requires medical professionals to record cases of infectious disease, but my belief is there is not a single country anywhere in the world where the system as a whole works as it is supposed to. I think it can and does work in certain limited instances where a disease is easy to diagnose and seen by everyone as vital to record. But I am hard pressed to think of an example where even this happens in a fool proof way.

We should not be surprised at this state of affairs. When national immunisation programmes were introduced it was not on the basis of a doubtful intervention that needed monitoring, but an overwhelmingly beneficial action that was so positive there was no need for monitoring and there was no expectation of it. Only with the development of more sophisticated civic organisation later on came the expectation of quantitative evaluation of such things. The monitoring methods were later add-ons not necessarily appropriate for the disease in question.

There are numerous real-life problems that get in the way of recording systems working. The main one is that almost invariably doctors and nurses are working all of the time trying to diagnose and treat people. That is what they feel is their purpose. Time spent writing and recording is hardly ever of direct and immediate benefit to their patients, so the more they do the less time they have to fulfil what they see as their primary role. Notification becomes a chore. Exceptions may occur in some well-developed countries where accurate recording is given a higher priority, but it will be mainly for medico-legal reasons. I don't think it happens in most countries.

D. Jenkinson, *Outbreak in the Village*, Springer Biographies, https://doi.org/10.1007/978-3-030-45485-2_16

In some parts of the world, if an infectious disease is diagnosed, there is a requirement to trace contacts, making a lot more work and acting as a disincentive to make a positive diagnosis.

In many instances a disease may occur (frequently with pertussis) when the doctor or nurse has provided immunisation against that disease. That makes the procedure look to have been a waste of time, and sometimes also money, creating negative feelings about a medical service that is trying its best to do a good job and keen to avoid negativity, so the disease is denied or called something else.

In many situations a positive diagnosis will only be made if there is laboratory confirmation, but often such facilities are not available, or only possible if the patient pays for it in a situation of relative poverty and is therefore omitted.

For some diseases there are backup data provided by laboratories where diseases are confirmed, but not all laboratories may be registered or required or able to produce numbers. This too will vary from country to country and also within countries.

In many countries that theoretically do proper recording, nobody collects the figures efficiently, or does it after so much delay that the collection seems pointless, so nobody bothers much. There are other countries where communication is so difficult that real physical problems prevent proper data gathering. Even some well-developed countries collect data differently in different states or regions, rendering comparisons awkward to say the least. France, for example, only collects pertussis data by sampling a small number of hospitals. Furthermore, if the gathered figures are not used properly, are incomplete or seen to be inaccurate, nobody will bother to take the process seriously and it is likely to degrade even further.

What we would really like to know is what is happening on a worldwide basis and to compare countries one with another to try to identify factors that could indicate success or failure in efforts to control infectious diseases to improve strategies.

In the case of pertussis, comparisons are enormously difficult between countries because each will have a different programme of immunisation, different numbers of doses, different ages at which they are given, and using different types of vaccine. Not only between countries but within the same country, dosage schedules and vaccine types change over time. The UK is a good example, changing to an accelerated schedule in 1990, introducing a booster in 2001 and changing to acellular vaccine in 2004.

But in the case of pertussis it is primarily the difficulty of making a clinical diagnosis and the lack of reliable testing methods until recently, and the variability with which testing has been used, that make differences between countries difficult to interpret. It is little wonder the world was slow to understand the changes that seemed to be happening to pertussis incidence in the USA and Australia in the mid and late noughties.

These two countries along with the UK are examples of countries where the populations are large and reporting systems for pertussis are believed to be consistent although not comprehensive. The USA was reporting increased cases of pertussis from 2004 in young infants and adolescents and the numbers kept on rising and reached a peak in 2012. Australia was reporting increased numbers from 2008 to 2012, particularly in the under ten-year-olds. The USA had moved to acellular vaccine in 1997 and Australia in 1999. Australia had also ceased an 18-month booster dose in 2003 and added it in teens instead. Evidence was accumulating that the rise in cases might be a result of acellular vaccine not being as affective as whole cell vaccine.

Canada is another country with similar practices and has used acellular vaccine since 1997. No significant resurgence has been seen. Likewise Denmark that is unique in using a vaccine with only a pertussis toxin component since 2004 and has not seen a resurgence.

The World Health Organization has a Strategic Advisory Group of Experts (*SAGE*) that advises on immunisation policy. *SAGE* reported in 2014 on a review of the evidence of the roles of acellular and whole cell vaccine in the 'resurgence' around the world [1]. Unsurprisingly the variation between the countries examined was enormous and inconsistent even though they were selected for the quality of their recording systems. Among several conclusions reached on the topic was the fact that resurgence had not been observed in countries that continued to use whole cell vaccine.

Evidence was also coming from a different direction, that of mathematical modelling, that could make predictions about pertussis incidence and explain current patterns. The National Infection Service of Public Health England produced such a paper in 2016 [2]. The main finding was that immunity from natural infection or whole cell vaccine lasted about 15 years, while that of acellular vaccine might be as little as five years. Therefore increased incidence could be expected. The same model predicted that an adolescent booster would reduce the incidence in that age group but would have little impact on the infant infection rate. Such modelling can be extremely valuable, but the accuracy of the prediction will reflect the accuracy of the incidence data on which it is based. I think that is inaccurate, as I have discussed in previous chapters.

Whatever the incidence data seem to be telling us, the accumulating evidence is strong that the acellular vaccines currently in use are inferior to whole cell vaccine and therefore the number of cases is likely to be higher in the acellular era than the whole cell era. Towards the end of the noughties experts in the UK were bracing themselves for a possible increase in pertussis confirmations that eventually came. For my own part I did not altogether accept the interpretation because my personal experience was telling me that any increase could be explained by increased recognition and confirmation. The Keyworth figures to 2018 have not shown any resurgence. My mind remains open about what exactly is going on.

Figure 16.1 illustrates the issue by comparing recent incidence statistics from Australia, the USA and England and Wales with the Keyworth incidence. Although all these three countries describe a resurgence, there is an enormous difference between the Australian high rate, which is numerically greater but most similar to Keyworth, and the other two which are much lower. Different rates of notification could be one reason. Different testing methods could be another. Changed immunisation schedule in Australia yet another. Australia has also used PCR testing more than the other countries, a test that detects subclinical and asymptomatic cases more easily. Making sense of it all is a big challenge and cannot yet be done.

*****

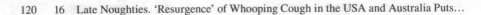

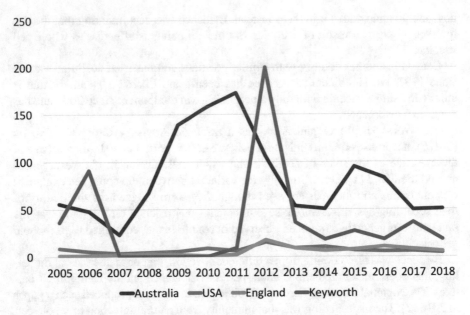

Fig. 16.1 Whooping cough cases in three countries and Keyworth, 2005–2018, as number per 100,000 population. (Data from PHE, CDC, Public Health Australia)

References

1. www.who.int/immunization/sage/meetings/2014/april/1_Pertussis_background_FINAL4_web.pdf
2. Choi YH, Campbell H, Amirthalingam G, Van Hoek AJ, Miller E. Investigating the pertussis resurgence in England and Wales, and options for future control. BMC Med. 2016 Dec;14(1):121.

Chapter 17
2012. Whooping Cough Diagnoses Peak in the UK But Remain Unchanged in Keyworth

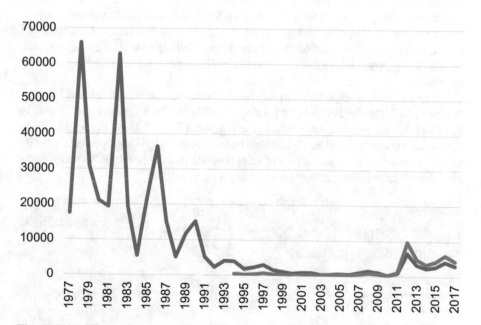

Fig. 17.1 Notifications of whooping cough 1977–2017 (blue) and laboratory confirmed cases 1994–2017 (brown) in England (and Wales to 2012) illustrating the increase that started in 2011

The full official whooping cough notification and confirmation panorama for the 40-year duration of this study is shown in Fig. 17.1. A total of 1121 laboratory confirmed cases in England and Wales in 2011 was roughly double the previous year's total and the rise accelerated in the latter half of that year in line with the four yearly cycles that would be expected to peak in 2012. I had not seen any cases in Keyworth between 2007 and 2010 but nationally there had been a small four yearly rise in 2008 with 1512 notifications.

When the rise started I felt pleased that at last the true situation was being recognised, namely that there was a lot of it about, always had been, and it was mainly in teens and adults, but as time went by I slowly realised this was not the way it was being interpreted. It was seen as a new phenomenon, as an *actual* rise in the number of cases. Although I could not be sure there was not a true rise, the fact remained that even with that rise, the diagnosis rate in Keyworth was still greater than the rest of the UK. And our 2012 peak was the lowest for 25 years!

Still focussed on trying to make people see that whooping cough had been there all the time, I was stirred by a news item in the *BMJ* on 23rd July 2012, reporting that the Health Protection Agency[1] had written to GPs alerting them to a nearly tenfold rise in pertussis cases in 15-year-olds and older since 2008. GPs should therefore be on the lookout for them and treat them with antibiotics where appropriate to reduce the risk of spread to infants. I had not seen this letter at the time as I had retired from active practice the previous year, although I was still keeping track of newly diagnosed whooping cough cases in Keyworth and interviewing them by telephone.

According to my experience in Keyworth there had *not* been a tenfold increase in cases at all, they had been there for decades (Fig. 15.2). So once again I wrote to the *BMJ*, which once again published my graph (Fig. 17.2) of the incidence in Keyworth versus England and Wales that demonstrated continuous peaks of activity of the disease that were way above official numbers for the previous 25 years, and I also described the high proportion in teenagers and adults [1].

INCREASE IN PERTUSSIS

May be due to increased recognition and diagnosis

I have meticulously studied all cases of whooping cough in my practice since 1977 and I have published many papers on the subject, mainly in the *BMJ*.

The increase in reported cases coincides with similar increases in the US and Australia.[1] A large proportion of this increase is probably the result of better recognition and diagnosis. My data provide good evidence that it never went away. What went away was the ability of doctors to recognise it, and in the absence of a practicable diagnostic test, official figures fell.

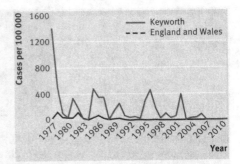

Numbers of cases of pertussis in Keyworth and England and Wales per 100 000 population (rounded to the nearest whole number), 1977-2011

Fig. 17.2 Whooping cough incidence 1977–2011, Keyworth and England and Wales. As published in the *BMJ* 25th August 2012

---

[1] Soon to be renamed Public Health England in 2013.

Experts had been watching the pertussis numbers closely because of acellular vaccine doubts. There was also evidence that although it prevented an individual getting ill from *Bordetella pertussis* infection, it could still allow the bacteria to multiply in the respiratory passages, giving the potential to pass it on to others. The same could happen to some degree with whole cell vaccine, but tests in baboons had shown the effect to be considerably worse with acellular vaccine. It was no great surprise therefore when case numbers rose, and it was presumably thought a better explanation than cases simply having been previously missed.

The Health Protection Agency certainly accepted that there was an element of increased recognition and confirmation, but made no attempt to quantify it as far as I could see. They may well have had a pragmatic reason for accepting the figures, because they were about to introduce a new pertussis booster in pregnancy, and a recent increase in the number of cases would help the uptake of the new booster dose and justify its introduction to a possibly hesitant clientele.

This new pertussis booster dose was introduced in the UK in October 2012 having been previously introduced successfully in the USA. It was groundbreaking and unexpected. It was easy, simple and effective. After 80 years of pertussis vaccine use, somebody tried this genius step that is turning out to be so effective in bridging that gap of four months after birth, when infants are so at risk from the worst effects of pertussis. They are susceptible because unlike some common infections, protective antibodies do not cross the placenta to the baby in sufficient quantity to prevent the disease. It had been discovered that a booster dose in later pregnancy, up to three weeks before delivery, would give the baby temporary immunity until it had had its own immunisation, which started at eight weeks and was completed at 16 weeks. Between the 16th and 32nd weeks of pregnancy is now considered the optimum time for the booster, although other countries differ slightly. Evaluation has shown it to be at least 90% effective and so it has been continued and adopted by many other countries.

The effect shows in the numbers of infants under four months catching pertussis and in the number of deaths from it. The benefits are now showing in the latest figures (Appendix 3). It could turn out to be a landmark change that gets closer to achieving the main objective of the vaccine, which is to stop babies dying of it, than any other change since the vaccine was introduced.

Whooping cough numbers continued to rise nationally in 2011 and 2012 when the peak of 9741 laboratory confirmations was reached, thereafter dropping and levelling off at around 4000 a year, about eight times the level of the doldrum years after the millennium. The overall pattern change shows clearly in Fig. 17.3 from Public Health England.

I, of course, believed it had never gone away, but this looked like a kind of rehabilitation and it is the right two thirds of the chart that show it. The elevated baseline from 2012 shows the constant background cases that are being recognised for what they are, the persistent circulating reservoir, which is mainly in teenagers and adults. The low left third of the chart reflects the diagnostic paralysis that I consider goes back to the mid-1990s. It might look like a big change on the chart, and so it is, but

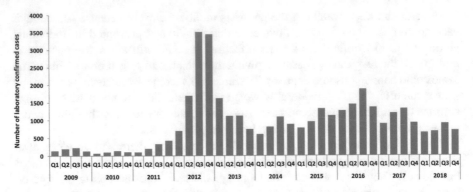

Fig. 17.3 Laboratory confirmed cases of pertussis by quarter in England 2008–2017 (Public Health England)

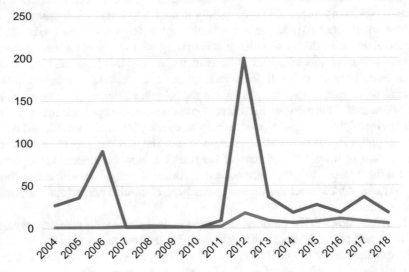

Fig. 17.4 Laboratory confirmed pertussis in England (plus Wales until 2012) (brown) and Keyworth cases (blue) 2004–2018 per 100,000 population

it is still only the tip of the iceberg. Even the peak number of cases in 2012 is only about a tenth of the Keyworth diagnostic rate (Fig. 17.4).

Although the 2012 rise was considerable compared with the previous two decades, it was small compared with the numbers seen in the 1980s. Figure 17.1 shows how it compares and includes the graph of the now dominating laboratory confirmed case numbers (brown), as opposed to notifications (blue).

Table 17.1 shows the age group structure of confirmed cases from 2011 to 2018 represented in Fig. 17.3.

Table 17.1 Laboratory confirmed cases of *Bordetella pertussis* in England 2011 to 2018 broken down by age group. Data from Public Health England

Age	<3 m	2–5 m	6–11 m	1–4y	5–9y	10–14y	15 + y	Total
No.	1203	286	135	580	1280	3297	29,073	35,850
%	3.4%	0.8%	0.38%	1.62%	3.6%	9.2%	81%	100%

Over four fifths are in teens and adults.

The level of 3.4% in under three-month-olds will probably strike some people as being high, but these are the most susceptible individuals. If this group is looked at in more detail it is even more disturbing, because their incidence expressed as a proportion of all under three-month-old babies in the population shows they have the highest *rate* of infection. Roughly one in every thousand is affected, although the very latest figures show a welcome drop (Appendix 3).

I believe the number of deaths from pertussis tells us something about the general incidence of the disease, and it once again suggests how inaccurate the official numbers may have been. Because unimmunised young babies are the most susceptible, they may behave like a barometer for the amount of *Bordetella pertussis* circulating in the general population. This will show in the number of under three-month-old babies with it, but probably more closely in the death rate, because some non-fatal cases may be missed. It can be missed as a cause of death too, but is less likely since reliable testing came along around 2002.

There were 35 deaths in the years 2001 to 2010, an average of 3.5 per year. There were 5 in 2011 and 14 in 2012 during that peak, a great increase. From 2013 to 2018 the average was 3.5 again, but meantime the pregnancy booster was available to affect the rate from 2013 onwards. The pregnancy booster is very effective, so without it the death rate could have been about double, seven a year, because although it is 90% effective the uptake was only about 65%. This is possibly confirming that there has been more pertussis about and may continue to circulate at a higher level than before. If this is so, the pregnancy booster is going to be a vital and key part of the future strategy against this disease.

Figure 17.4 shows the recent pattern in Keyworth compared with the rest of the country.

The graph shows the absence of correspondence between Keyworth and England and Wales pre-2010, but after that there is a relationship again which is a roughly five to tenfold difference, much the same as in the 1970s and 1980s when this story started.

I have put considerable thought into reconciling the fact that the Keyworth figures have not really changed, with a possible resurgence for which there seems a mixed bag of evidence. Is it possible for there to be more pertussis infections about without the number of clinically apparent cases changing? One explanation could be if subclinical cases that do not reach a threshold of three weeks of paroxysmal coughing were more infectious if previously immunised with acellular vaccine. They might be more likely to pass it on because it is suspected to be poorer at preventing transmission than whole cell vaccine.

Figure 17.5 diagrammatically represents my speculation about how whole cell vaccine might be compared with acellular vaccine in the way different severities are

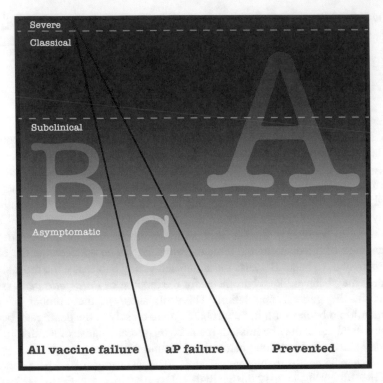

Fig. 17.5 A representation of how different pertussis vaccines might work across the whole spectrum of *Bordetella pertussis* infections

prevented or not. The square represents the full spectrum of *Bordetella pertussis* infections, graded from severe at the top, through subclinical, to asymptomatic at the bottom. Area 'A' plus 'C' represents infections prevented by whole cell vaccine. Area 'A' represents the infections prevented by acellular vaccine. Area 'B' represents all vaccine failure. Area 'C' represents infections that occur after acellular vaccination that would not have occurred after whole cell vaccination.

Area 'C' shows how acellular vaccine may allow the escape of more subclinical and asymptomatic cases than severe cases. The concept is not directly evidence based and the proportions cannot be considered quantitative. It has been calculated, however, that for each clinically identifiable case of pertussis there are between 5 and 20 below that threshold. It also illustrates one of the most important aspects of all pertussis vaccines that is little appreciated; that being that they prevent severe whooping cough better than mild whooping cough.

*****

Reference

1. Jenkinson D. Increase in pertussis may be due to increased recognition and diagnosis. Letter. BMJ. 2012;345:31.

Chapter 18
What Lies Ahead?

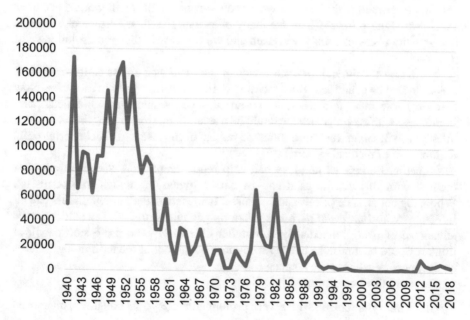

Fig. 18.1 Whooping cough in England (and Wales to 2012) 1940–2018. Notifications to 2009, then laboratory confirmed cases

Figure 18.1 shows the full incidence history of whooping cough in England and Wales measured by notifications and confirmations from when it was first notifiable in 1940. Compare the alps on the left with the anthill of our current situation on the right. That is what pertussis immunisation has achieved. The hump in the middle is what caused this book to be written!

Whooping cough vaccine has saved millions of lives worldwide since it was introduced in the 1940s, but it is still an area of challenge in infectious disease control. Originally thought to be a childhood disease that gave long-lasting immunity, we now know repeated unnoticed infections occur throughout life and boost immunity. Immunisation strategies have to take account of this, but detailed understanding is lacking.

Now that we know around 10% of persistent non-feverish acute coughing illnesses in teens and adults are likely to be caused by *Bordetella pertussis*, will patients and healthcare professionals who are aware of this expect laboratory testing for it? If so, the number of confirmations will increase enormously, rendering comparisons year by year almost meaningless, but it would bring home the point of the need to protect the most vulnerable. Currently the pregnancy booster looks best for this. The success of this booster is based on the fact that everyone who currently gets it was primed with the older whole cell vaccine. Will it still work if acellular vaccine was the primer? There are many questions like this. Fortunately, there is high-quality research being undertaken and we can expect that good solutions will be found.

It will be very difficult to track trends in the incidence unless confirmed infections can also be quantified with a measure of the clinical severity in each case. Now that we know there is a whole spectrum of severity, and that subclinical cases immunised with acellular vaccine might be transmitters, simply counting confirmed infections will tell us very little about the impact of the disease in social terms or its requirement of medical resources.

Sometimes a new terminology can help reduce confusion. We already have a conflict with this disease as it can be called 'whooping cough' or 'pertussis'. Whooping cough is descriptive and pertussis is causative. Perhaps we should call all infections with *Bordetella pertussis* 'whoopingcough-pertussis'. It would have the advantage of being immediately comprehensible to lay and profession people. I think it would be a mistake to drop whooping cough out of the terminology as it is so deeply entrenched in the language.

There are three categories of infection:

1. Classical: symptomatic with severe prolonged paroxysmal coughing, diagnosed clinically.
2. Subclinical: diagnosed by laboratory test.
3. Asymptomatic: diagnosed serendipitously or by survey investigation.

All cases need to have the category recorded as one of the above if statistics are going to be meaningful in the future. Classical whooping cough has a distinct meaning in everyday life with important consequences and needs to be recorded as such. The other kinds of *Bordetella pertussis* infection may be important clinically, and they certainly are from an epidemiological angle, so should be recorded thus as subclinical or asymptomatic.

Technology can rapidly change the way we deal with problems, and there is a new development that could dramatically affect the way we deal with pertussis. A 'point of care' biochip test has been developed that detects pertussis DNA and can be done easily [1]. It gives a result within an hour without special equipment. It is inexpensive, and no doubt others are on the way which could be even quicker. It would enable the presence of *Bordetella pertussis* to be detected whatever the symptoms might or might not be. It would be most useful if pertussis was merely suspected, perhaps early on in the infection process when an antibiotic and quarantine might be useful. It would, however, play havoc with the disease statistics the way they are currently recorded.

This may already be occurring in Australia and the USA where PCR is more commonly used. Research has shown that because the PCR is positive right from the early stages of infection, many never go on to develop clinical whooping cough. These cases elevate the statistics and may be feeding belief in resurgence. That is why it is as important to record when a clinical diagnosis threshold is reached as it is to test for the bacteria.

There will also need to be a better understanding of the whole disease among medical and lay people if confidence in the currently used vaccine is to be maintained. This could be lost if the bare facts are taken out of context. The paradox needs to be well understood that even though the *individual* immunity afforded may be relatively short and incomplete, the protection of vulnerable infants and others through *herd* immunity and reduced severity in the immunised is great, even with acellular vaccine.

Anyone who reads this book and is 'vaccine-hesitant' will, I hope, come away understanding that vaccines are only worth universally recommending if they are certain to do vastly more good than harm, and that the processes that determine how much good they do are extremely complex and therefore have to be accepted with a high degree of trust. Some people find that trust more difficult than others. In most circumstances where vaccine acceptance has fallen to dangerously low levels the official response had been to emphasise the benefits of the vaccine and the harm from omitting it, the most logical response no doubt. Unfortunately it does not address the main problem, which is lack of trust. Medical authorities reason that it is they whom people trust most with medical issues, ignoring the fact that medical experts often disagree, and that they might have a vested interest. If a trusted celebrity such as David Attenborough declared it a good thing, which he no doubt would, I would predict a much higher acceptance rate. That might be a better way forward. Compulsion is unlikely to work in our relatively free society.

A recent review of 'pertussis and pertussis vaccine mistakes over the last 112 years' by an internationally acknowledged expert made a strong case for development of a new vaccine [2]. Such is the doubt about acellular vaccine that the *WHO* advises countries that have not changed, to stick with the whole cell type, which with hindsight was a lot better than we thought. Whole cell vaccine is not however a standard product, and some are undoubtedly better than others. The apparent effectiveness of acellular vaccines resulted in part from comparisons with poor whole cell types. The best whole cell vaccines probably give similar protection to the natural infection.

Another big question that needs to be considered is what happens when acellular vaccine fails to protect. In that situation *Bordetella pertussis* gets a hold, and that would normally be expected to generate a full natural immunity for a period of time, but that might be reduced by previous immunisation with acellular vaccine. We know that acellular vaccine makes it more possible for the organism to survive in the victim for longer than after whole cell vaccine, but the extent to which this phenomenon reflects the immune response and whether it is clinically important is not known.

Bordetella pertussis itself has changed genetically in several ways. This is not a surprise, but it has changed more where acellular vaccine has been used. Infectious diseases have frequently been subject to change. It is in the nature of such things. Long before DNA technology could reveal the detail, we knew that the organism could be typed according to the presence of different antigens on its surface. Some combinations seemed to be associated with more severe disease. There was a tendency for the predominant organism to change to a different type if the vaccine was effective at preventing spread of the dominant circulating variety.

Some of the observed genetic changes include small differences in pertussis toxin, and in other places, absence of pertactin adhesin production, a component of many acellular vaccines. Although the changes have been noticed, it has not been possible to link them to 'resurgence', but they show there is potential for it, and it seems that more genetic variation has developed since the introduction of acellular vaccine.

Resistance to erythromycin, the antibiotic used against *Bordetella pertussis*, has been detected recently in some countries.

Nobody yet knows what level of antibodies are protective, or in what combination, so vaccine effectiveness measurement is still only possible by looking at what happens in real life situations until we know more about the nature of the immunity and other defences against *Bordetella pertussis*. It uses many tricks to invade our bodies, one of them being the ability to hide from our immune system.

A lot of work is going on to produce a new good *Bordetella pertussis* vaccine in spite of our lack of understanding. The process of developing vaccines against bacteria has always been a severe challenge. The really effective and successful vaccines are mainly against viruses such as measles and polio. The ideal vaccine should kill the bacteria as soon as they arrive in our bodies, so they get no chance to spread to others. It should also produce better and longer protection than the current whole cell vaccine and of course be free from adverse reactions. That is a big challenge.

Researchers are looking to see if removing the elements that cause the unpleasant reactions from the old vaccine can provide a solution. Another way is by changing the adjuvant in the vaccines. Adjuvants improve the immune response. Alum is currently used. Organic adjuvants are known that can work better.

Genetically modified vaccine is another possibility and is achieved by leaving the basic bacterium intact with all the bits that are needed for our immune system to recognise it, but disabling its ability to damage our cells. Such a vaccine could be live and could not only be effective but could be administered in the form of nose drops. There is already such a vaccine under test in human volunteers [3]. It has the name BPZE1. But vaccine development and licencing take a long time. It would take many years for it or a similar one to replace the current vaccine even if it fulfilled all the hopes for it.

There is yet another string to the bow of prevention which may work its way into pertussis vaccine technology. The concept arises from the fact that some people catch it and some people don't, but because of something other than immunity. Sweden stopped pertussis immunisation in 1986 and didn't restart for ten years, providing valuable information about the natural infection in a modern society. Most unimmunised children escaped overt whooping cough.

A high proportion of unimmunised household contacts who have not had the disease catch it, but not all. Most people in casual contact don't appear to get it. This implies that you need a high bacterial load to get infected. Why is this? To reliably infect a human experimentally requires 100,000 bacteria![1] You could say that it is just bodily defences working well, but experience from other fields tells us our microbiome could be contributing. Microbiome is the name given to the bacterial life permanently in and on us. These bacteria often outnumber our own cells and their importance to our well-being is being increasingly recognised. Our nose and throat are the home of friendly bacteria that live on our cellular debris and secretions and each other and exist harmoniously with our tissue cells. Their presence will inhibit invaders. A disturbed microbiome is unlikely to repel so well. Perhaps a change in our microbiome can make the difference between *Bordetella pertussis* taking a weak hold or a strong hold, or none at all.

What can affect our microbiome? Antibiotics? Diet? Other infections? Chance? All these things and many others. The understanding of our gut microbiome, which has been studied better, tells us that. Gastroenterologist sometimes do faecal transplants to restore a deranged gut microbiome in the case of infection with *Clostridium difficile* and other conditions such as irritable bowel syndrome. Our nose could be the same, and knowing how our friendly nasal bacteria might repel *Bordetella pertussis* could help our understanding and inform vaccine development.

A question that I find particularly intriguing is why there is such a range of severity. I think I have seen clues while I have spent 40 years observing this disease, but it is mostly speculation. I feel it is possible that in order to cause a significant problem, *Bordetella pertussis* may have to work in partnership with some other infection. I think a recent or concurrent viral upper respiratory infection may play a part in the ability of *Bordetella pertussis* to establish itself and cause symptoms of whooping cough or possibly sometimes even depend on it. I have seen some evidence for this.

For example, I have found that index cases in a household are more severe than subsequent ones. I have also found that the index case in a household is more likely to have attended a GP in previous weeks for a respiratory infection than subsequent cases in the household. I have found that only a third of a sample of cases remember catarrhal symptoms, which were perhaps a viral cold rather than the pertussis. Perhaps the little explosive outbreaks observed are caused by the introducer spreading a virus and pertussis simultaneously; the differences in incubation periods

---

[1] Actually it is even more than this because it is 100,000 'colony forming units'. A unit being a single bacterium or a whole clump.

between the two will soon divide the dual spread and so it fizzles out. Viral infections are frequently found alongside *Bordetella pertussis,* and for a long time caused people to believe that whooping cough could be caused by some viruses. Perhaps if there is such an association between *Bordetella pertussis* and a helper microorganism, the search could be narrowed by looking for a micro-organism that has a preference for females. There could even be several organisms involved. The unusual sex ratio of this unique cough raises the possibility of it being an autoimmune phenomenon, which females are more prone to.

What will the path to better immunisation be? I have no special knowledge and I can only speculate, but I don't think there is anything just around the corner. The introduction of the pregnancy booster can deal a hammer blow to infantile pertussis if full advantage is taken of it. It is possible that changes to that vaccine's components might make it even better. The fact that it is a combined vaccine with tetanus, diphtheria and polio is a disadvantage. I hope work is going on to produce a monovalent pertussis vaccine without the other two. Manufacturers may need to work more closely with researchers and government agencies to overcome the bureaucratic obstacles that often unintentionally get in the way of progress of this straightforward kind.

It has recently been discovered that baboons can be infected deliberately with *Bordetella pertussis* and they seem to become ill from it in much the same way as humans. Some investigations have been done into acellular vaccine in these animals. This is how we learnt that acellular vaccine can prevent them getting whooping cough but still allow *Bordetella pertussis* to multiply in the tissue of the nose and throat.

Baboon experiments have proved that *Bordetella pertussis* is transmitted by droplets through the air. As if we didn't know already! But it illustrates how little we are really sure about with this disease, and if we are going to make progress in understanding it, we have to know all the details.

Fortunately, progress with vaccines can be made experimentally without understanding the whys and wherefores, but it is an extremely prolonged and expensive process. Although the susceptibility of baboons allows greater scope for investigation of the disease, there are still strict ethical limits to how far it can go, and the findings may well not apply to humans. But if it turns out that we react the same way as baboons it could be shortcut to a better vaccine.

The burden of pertussis in adults is now better recognised and ways of reducing it are being researched. Some countries are advising a booster every 10 years. It is not thought it would have much impact, one reason being the natural background boosting that goes on. A theoretical solution to this problem would be to give a booster more frequently, possibly every three years, but that is unlikely to be acceptable. It could perhaps be considered as a solution for selected vulnerable groups, but vaccines are not currently licensed to be used that frequently anyway. Once again, greater co-operation between the agencies involved would help making appropriate moves at a time of need easier.

Whatever the next vaccine looks like will depend on its effectiveness in clinical trials in the context of a greater understanding of the organism itself. At present it seems there are still more unknowns than knowns.

In theory whoopingcough-pertussis can be eradicated. The technological steps necessary to reach that goal are understood already. Incentivising those who can make it happen, proving it can be done safely, and gaining the consent of humanity to accept it are the potential obstacles in the way.

*****

I remain grounded by my research in the Keyworth practice, tiny though it is, which has kept count of cases based on clinical diagnosis, not laboratory diagnosis, for over 40 years. We have not seen a resurgence and are still diagnosing approximately three to ten times more (Fig. 17.4) than are confirmed in the rest of England, the same as we were doing 40 years ago.

Counting clinical cases of whooping cough as well as laboratory confirmations might bring some enlightenment. However, the large increase in numbers of unimmunised babies with pertussis in 2012 (now fortunately preventable with the pregnancy booster) may be telling us there is more *Bordetella pertussis* circulating than there used to be and either way it is going to remain a serious threat. I have stated my suspicion that increased subclinical cases might explain some of the enigma.

The Keyworth Medical Practice will endeavour to remain active in diagnosing clinical whooping cough based on three weeks of paroxysmal coughing just as before. It is just possible it is the only practice in the world with such continuity and we think this might remain useful.

The pace of change in this field is rapid. Research mainly comes out of academic and industrial institutions. I firmly believe that observations in general practice can also contribute in this broad and complex field.

*****

References

1. M, Macias N, Shen F, Bard JD, Domínguez DC, Li X. Rapid and accurate diagnosis of the respiratory disease pertussis on a point-of-care biochip. EClinicalMedicine. 2019 Mar 5;
2. Cherry JD. The 112-year odyssey of pertussis and pertussis vaccines—mistakes made and implications for the future. Journal of the Pediatric Infectious Diseases Society. 2019;10:1093.
3. Thorstensson R, Trollfors B, Al-Tawil N, Jahnmatz M, Bergström J, Ljungman M, et al. A phase I clinical study of a live attenuated Bordetella pertussis vaccine-BPZE1; a single Centre, double-blind, placebo-controlled, dose-escalating study of BPZE1 given intranasally to healthy adult male volunteers. PLoS One. 2014;9(1):e83449.

Chapter 19
Epilogue

A happy but unexpected event involving whooping cough occurred in May 2013. I have always been fascinated by the history of the American West, and in that month I drove the route of the Oregon Trail from Independence, Missouri, to Portland, Oregon. Between 1846 to 1869, half a million migrants followed this 2000 mile route to find a better life. Many relics and mementos of the migration remain today in the form of museums, wells, springs, and deeply rutted, now abandoned tracks in the wildernesses of the mid-west, often marked with information boards telling epic tales. These places are often remote, forgotten and difficult to access. There are crude graveyards marking the last resting places of the one in ten who died on the journey.

Most migrants were heading for the Willamette Valley, a vast fertile area for successful farming. They left Missouri in springtime when the grass started to grow, and they had to pass through the Rockies before the first snow. They had six months to do it in or they were doomed to freeze.

I was making the journey by car in three weeks. I needed to be back home a few weeks later to make a trip to Switzerland to research a nineteenth-century artist I had been told I was distantly related to called Eugène Burnand.

In the second week on the Oregon Trail I received a desperate email from the mother of 7-week-old baby with possible whooping cough. I hardly ever received inquiries about babies. This was very unusual. I reproduce the email exchange below with permission. It beautifully illustrates the fright engendered by a baby with whooping cough, and from a personal point of view I somehow felt it embedded me into the legacy of the early pioneers whose trail I had been following.

© The Editor(s) (if applicable) and The Author(s), under exclusive license to
Springer Nature Switzerland AG 2020
D. Jenkinson, *Outbreak in the Village*, Springer Biographies,
https://doi.org/10.1007/978-3-030-45485-2_19

May 13th 2013.
Dear Dr. Jenkinson,

We live in Eugene, Oregon, USA, Can you help us, please? Our baby was born March 25th, just seven weeks ago. On May 6th she started coughing in fits, with no other symptoms. May 8th she saw a doctor and was diagnosed with bronchiolitis. Her temperature was 100.4 F. May 10th she started whooping on inhale during coughing fits. Today (May 13th) the doctor did a nasopharyngeal swab for whooping cough. They say they will get the results tomorrow morning. They say that if she has a positive test result, they will prescribe antibiotics. Her temperature today was 99.2 degrees F.

I would very much like to avoid antibiotics, especially with such a young baby who is still establishing healthy intestinal flora. I have read that antibiotics do not help with whooping cough unless prescribed at the very beginning of the disease (they would make her less contagious, but not shorten or improve her symptoms). Would you please tell me whether you think antibiotics will help my baby?

In Eugene, where we live, there is a high incidence of vaccine avoidance. Whooping cough is pretty common here. Our baby is healthy and big for her age. She weighs currently 12 US pounds 13 ounces (7 ounces more than last Wednesday), is not dehydrated, is breast-fed and nurses well, has no congestion and can thus nurse easily. She sleeps well and is happy when awake, and has no signs of respiratory distress. Our doctor says her lungs sound clear. Her only symptom remains the coughing fits, with whooping on inhale, and occasionally vomiting at the end of the cough. She is getting more uncomfortable with the cough, and fusses at the end of a coughing fit until she is comforted. She is coughing up clear-to-white sticky mucus, that stretches like egg white between my fingers. Sometimes she chokes on it as part of the coughing. She has between one and five coughing fits per hour, with no fits at all for about half of the hours of each day.

She sleeps with us, and I am able to respond to her quickly as a result. I feel capable of diagnosing respiratory distress (I am in nursing school), and can help her long enough to get her to a hospital if needed. Is there anything I need to know about helping a baby of her age to recover quickly? (Other than the usual: nurse, rest, take care of me, stay hydrated.) It is hard to find good information about whooping cough in a baby of her age, because mostly it is scary stories about hospital stays, IVs, and intubation.

If you are able, please tell me whether you think she would benefit from antibiotics, and whether there is anything else we can do for her. Thank you very much.

Blessings
K

Dear K
Difficult to say whether whooping cough or not. Sounds fairly mild if it is. I would not hesitate to use antibiotics if advised. They can be lifesaving and the supposed disadvantages are exaggerated, trivial or mythical. Odd you are in Oregon. I

am on vacation following the route of the Oregon trail! At present I am in Kearney, Nebraska. I shall be in Oregon in about 10 days. It would be nice to meet you. Let me know how she gets on.

Dear Dr. Jenkinson,

Thank you for your response. After how we fared last night, I am ready to use antibiotics or anything else the doctor recommends. Our baby (her name is Integrity, by the way, and we call her Tegra) had several coughing fits, and in some of them it seemed like she didn't know how to breathe. She is now almost continuously clearing her throat, and breathing unevenly. I am calling the doctor as soon as their office opens.

I would be delighted to meet you when you get to Oregon! We live on the west edge of Eugene, near the Fern Ridge reservoir, very close to the Applegate Trail section of the Oregon Trail. Fern Ridge is the best place in Oregon to go birding, if you like that kind of thing.

I will let you know how Tegra fares.

Blessings,
K

Good luck. Keep in touch. Will hope to meet up
Doug

Hi Dr. Jenkinson,

Tegra's nasopharyngeal swab came back positive for pertussis. We are in the hospital overnight for monitoring. I feel we are in good hands here at the hospital; their information and protocol seem similar to what you recommend on your website. My doctor says he is thankful that I came in a second time to ask for a pertussis test.

I would love to see you when you come this way. Let me know when you are nearby. Our home is a 33-acre farm with an 1880s farm house and barn from the original homesteaders.

Blessings, and have a good trip.

I could hardly believe where they lived. They were in the town of Eugene (the same name as the artist in my next destination) which was in the Willamette Valley, Oregon! It was virtually my destination! I had to do a home visit and Tegra's parents had kindly invited me to their homestead. Tegra herself was considerably improved by the time I arrived and was only getting occasional paroxysms. I was made very welcome and I enjoyed a snack and a drink as I made friends with Tegra before getting on my way to Portland. We have all kept in touch ever since and Tegra is a thriving girl.

Fig. 19.1 The author with
baby Tegra who was
recovering from whooping
cough in Eugene,
Oregon, in 2013

A very satisfactory ending (Fig. 19.1).

Appendix 1: About Bordetella Bacteria

Bordetella pertussis is a species of bacterium in the Bordetella genus. Bacteria are single-celled organisms that were among the first life forms to evolve. They exist everywhere that other life exists, even in places with extreme conditions where other life cannot survive, such as in high temperatures or with usually toxic chemicals. The total bacterial biomass on the planet exceeds that of plants and animals. Little or nothing is known about most of them, there probably being between 10 million and a billion different species. They must not be confused with viruses, which are different organisms altogether, being even smaller, simpler in structure and only able to reproduce by hijacking the apparatus of cellular organisms.

Bacteria form an integral part of the balance of life on earth, mainly concerning themselves with living on the dead remains of more complex organisms. They are almost everywhere. There can be 40 million bacteria in a gram of soil. Many live on or in other plants and animals. An adult human consists of about 30 trillion cells. An equal number of bacterial cells or more live on our skin, in our mouth, nose, and principally in our gut. They are vital to our well-being, breaking down food, making essential chemicals, interacting with our immune system and warding off less friendly species. A healthy human body is an interaction between our organs and our microbiome.

We are generally unaware of the friendly bacteria we carry but we become very concerned with the unfriendly ones that can interfere with our bodily functions. They are the pathogenic bacteria and we have to find ways of fighting them if our immune system fails to do so. *Bordetella pertussis* is one of them. The usual method is by using antibiotics. These are usually successful and have completely changed our relationship with pathogenic bacteria. We have become very good at killing them, although they have developed ways of fighting the antibiotics currently in use and we are coming to realise the seriousness of this.

Pathogenic bacteria do not usually set out to kill us. If they do so they are likely to die themselves. Often their best strategy is to strike a balance between maximising the chance of getting passed on without doing the host too much harm.

In the case of *Bordetella pertussis* it is a battle of considerable complexity. Most of the lifestyle of *Bordetella pertussis* revolves around being passed between individuals who hardly notice their transit in most cases, whether in immunised or immunised populations. Causing the death of a small proportion of babies is an accidental and inessential part of their struggle to survive.

From the human perspective pertussis is the most lethal of childhood infections and has failed to be controlled by antibiotics or immunisation properly, so is now getting more research attention than ever, as the need for an improved vaccine becomes more urgent. Consequently, much is now known about *Bordetella pertussis,* but there are still many important questions awaiting answers. Such as the cause of the unique cough, and precisely why some people get symptoms of whooping cough when infected and not others.

Although there are 10 or more known species of *Bordetella*, apart from *Bordetella pertussis* only three others are in any way known to us as humans. Dog lovers and vets may be familiar with 'kennel cough', an infection that can be problematic in the way suggested by the name. One cause is *Bordetella bronchiseptica*, a species that affects not just dogs but many kinds of animal the world over. *Bordetella parapertussis* causes much the same symptoms as *pertussis* and turns up in a small proportion of cases in outbreaks of whooping cough.[1] It causes somewhat less severe symptoms than *pertussis*. *Bordetella holmesii* is the third and was only discovered in 1995. It also causes whooping cough symptom but is much less common and seems capable of causing more severe general illness. *Pertussis, parapertussis* and *holmesii* are the only *Bordetella* species that normally affect humans.

From genetic analysis it is possible to say that the three human species evolved from *bronchiseptica* or a similar ancestor of it. They are genetically simpler and were able to invade new territory, i.e. humans, by eliminating genes that aided invasion of other species but prevented them using a human host.

Bordetella pertussis is a very small oval-shaped bacterium with fastidious nutritional requirements. It can be grown on blood agar but takes five days for minute colonies to form. In order to reproduce in humans, *Bordetella* species need to access ciliated surface cells in the nose, throat or lower airways. Cilia are microscopic fronds on the surface of individual cells that are able to waft mucus in a particular direction, generally towards the mouth where it will be swallowed, down from the nose and up from the lungs. *Bordetella pertussis* bacteria reproduce in the mucus of that ciliated cell environment, but before they can do so they need to stick to the cilia. They achieve this by producing adhesins that help to keep them in place. Otherwise they would get wafted away, swallowed and destroyed. They would also get engulfed by cells of the immune system such as macrocytes that are on the lookout for foreign material. *Bordetella pertussis* has ways of hiding from the immune

---

[1] There is also a variety of parapertussis that only affects sheep.

system using toxins to disrupt cells of that system. Possibly for that reason, unlike most acute childhood infections, only mild fever is caused, and even then, it is only in the early stage, not during the paroxysmal phase, unless there is a complicating secondary bacterial infection.

Once established, *Bordetella pertussis* cells produce toxic substances that have a damaging effect on the cells they are attacking (white blood cells of the immune system for example), and they can kill those cells. This process may be complete within a week or two of the initial invasion, but even if all the bacteria are killed by the immune system or antibiotics, the toxins persist or will have done the damage that causes the symptoms to continue for many weeks longer. It is sometimes possible to detect live bacteria for up to six weeks after the start of symptoms, but it is usually extremely difficult after just three.

A rather unusual and intriguing finding is that the bacteria are able to invade alveolar[2] macrophages and survive for a time [1]. Intracellular survival has also been observed in leucocytes and cultured alveolar epithelial cells [2]. In the latter case it was shown that if the bacteria outside the cells are all killed off, the ones hidden inside the cells could reinvade the outside again. The way *Bordetella pertussis* bacteria evade normal immune responses may shed light eventually on why and how immunity is boosted from time to time without symptoms. The fact that intracellular survival can happen tempts the thought that such a phenomenon might be relevant to a fuller understanding of this disease.

The best known of the toxins produced is aptly named 'pertussis toxin' and is similar to the toxins that are produced by cholera and dysentery bacteria, and other types too. It is a complex chemical 'missile' that aids adhesion and colonisation, and also modifies the cellular immune responses. It is also responsible for the big increase in lymphocytes, a type of white blood cell involved in immunity, more commonly seen in viral as opposed to bacterial infections. Neutrophils, another type of white blood cell are also increased, and this toxin causes all these cells to form small clumps. These clumps can then block small arteries in the lungs, a process that is believed to happen in severely affected babies, causing raised blood pressure in the lungs, the consequences of which can be so lethal for them. The paradoxical thing about pertussis toxin, which sounds as if it ought to explain almost everything, is that only *Bordetella pertussis* produces it. *Parapertussis* and *holmesii* that cause similar symptoms don't, even though they carry the genetic code for it, as does *bronchiseptica*. Furthermore, pertussis toxin's manifestations only occur in naïve subjects, not in subsequent infections or in the immunised.

It has been discovered recently that *bronchiseptica* has a double life, being also able to live and reproduce within free-living amoebas, possibly acting as an environmental reservoir and host vector for the *bronchiseptica* infections that affect many animals [3]. This is an intriguing serendipitous discovery and generates tantalising

---

[2]Alveoli are the minute air sacs in the most distant parts of the lungs where gas exchange takes place.

thoughts about whether something similar could be discovered about the species that affect humans.

Other known toxins produced by *Bordetella pertussis* are adenylate cyclase toxin and dermonecrotic toxin. These also play a part in establishing the infection, and although their role is known in considerable detail from experiments in animals such as mice, precisely if and how these substances are deployed in the invasion of the human respiratory tract is not.

Several adhesins are produced that help the bacteria stick to airway cells. They include filamentous haemagglutinin and pertactin and are thought to be important antigens that help the immune system recognise the bacteria, so are often included as components of the acellular vaccine. That they do not work as well as whole cell vaccine is hardly surprising given that whole cell vaccine contains approximately 3000 different proteins, including the ones used in the purified versions. Even the natural infection doesn't produce long-lasting immunity, so the challenge to find a better one is formidable.

Possibly the biggest mystery still remains. What causes the choking cough that can go on for months? It is not sputum production. It is not inflammation. Still nobody knows. Calling it an enigmatic disease may be a bit of an understatement.

*****

References

1. Friedman RL, Nordensson K, Wilson L, Akporiaye ET, Yocum DE. Uptake and intracellular survival of Bordetella pertussis in human macrophages. Infection and immunity. 1992 Nov 1;60(11):4578–85.
2. Lamberti Y, Gorgojo J, Massillo C, Rodriguez ME. Bordetella pertussis entry into respiratory epithelial cells and intracellular survival. Pathogens and disease. 2013 Dec 1;69(3):194–204.
3. Taylor-Mulneix DL, Bendor L, Linz B, Rivera I, Ryman VE, Dewan KK, Wagner SM, Wilson EF, Hilburger LJ, Cuff LE, West CM. Bordetella bronchiseptica exploits the complex life cycle of Dictyostelium discoideum as an amplifying transmission vector. PLoS biology 2017;15(4):e2000420.

Appendix 2: Age Incidence Changes in Keyworth Whooping Cough Patients

Throughout the writing of this book I have been pulled in two directions by two sets of information. One set is the observations I have made and recorded in Keyworth on patients I have studied and got to know personally over about 40 years. The other comes from the many scientific papers that have been written, and I have read, that result from the work of other researchers. I, of course, as a scientist, accept the validity of the research that is done elsewhere, and I know that it is the way a fuller understanding of the disease will be achieved. At the same time, I know that small observations of detail, of the sort my study can make, can sometimes be important and give insights into deeper truths.

I am also very well aware of the errors that can occur in small studies like mine, when random associations can appear meaningful, when in fact they have arisen by chance alone, or because of the biases that are inevitable in any study. Scientists have now developed effective ways of combining different studies of the same thing in meta-analyses that can extract knowledge more accurately and these have become extremely valuable. I have worked at the other extreme and tried to extract what information I can from a comparatively small amount of data on the assumption that it might be true, because I cannot tell what is and what isn't. If my data contradicted known hard facts, I would discard it. If it does not, I have assumed it might be true, and kept it. Whether it turns out to be true in the long run will have to wait for time and other studies to determine.

There are two classes of data from Keyworth. One is hard data like age, and whether a swab is positive or negative, and immunisation status. The other is anecdotal, such as the observation that whooping cough relieves asthma. I have been careful to be clear about which category I might be discussing as I have gone along. Other researchers might analyse my data in different and cleverer ways. I am happy to share the database of my anonymised 744 patients with the scientific community.

Figure A2.1 shows how the average age of Keyworth whooping cough patients has changed over the study period. The sample intervals are not the same, because I

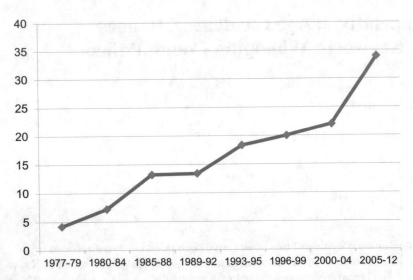

Fig. A2.1 Average age in years from 1977 to 2012 in Keyworth whooping cough cases

arbitrarily chose intervals that contained sufficient numbers to give a meaningful answer. I should also point out that average is only one way of describing a range of ages at any point, but I think it is the best for my purpose here as everyone knows what average means.[1]

There is a clear and almost steady rise over the 35 years from 4 years to 34 years. The reason for the change is undoubtedly the result of roughly half of children not receiving pertussis vaccine for about a decade after 1974, which meant that many of these children caught whooping cough at a young age. But there is also the introduction of a pertussis booster to four-year-olds in 2001, which could be expected to reduce the number of cases from that age upwards subsequently. We can no longer say the upward trend is a result of the population of older people with naturally acquired immunity getting smaller, because we now know natural immunity from having the disease is much the same as immunisation with whole cell vaccine.

When I break down the numbers by age (again by arbitrarily selected age groups), it is possible to make some general observations about trends in Keyworth over 35 years (Fig. A2.2).

The number of under ones (red) has gradually diminished to zero. This is because from about 2000, virtually all babies were having pertussis vaccine. Of course, under 4-month-olds are still susceptible but the Keyworth population is too small to represent this group.

The one- to nine-year-olds (green) are predominant in 1977 and 1978 because they are the susceptible unimmunised group at the time. They continue to feature

---

[1] Other expressions are 'median', which is the middle value, and mode, which is the most frequent value.

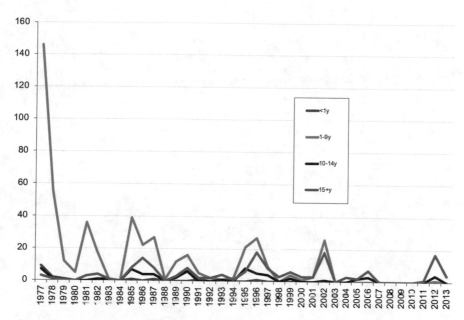

Fig. A2.2 Keyworth whooping cough cases by age group 1977–2013

strongly until 2003, then suddenly drop to a much lower level. That could be the result of the addition of pertussis to the preschool booster in 2001.

The 10- to 14-year age group (black) has remained fairly steady throughout. A small number of them could have had a booster dose.

The 15-plus age group (blue) grew for 10 years from 1977 and remained fairly steady thereafter. It is possible that in the early years I was failing to diagnose it in adults because I did not consider it at the time. I cannot say if that was so or not, but I am reluctant to think my mind was not open to recognising it because I identified three in 1977–8. With hindsight and a modern understanding of the probable epidemiology at the time, there perhaps ought to have been more. I cannot explain their absence unless natural immunity is better than whole cell vaccine after all.

Appendix 3: A Critical Look at Recent Pertussis Statistics in England

2011 and 2012 were landmark years for pertussis. The number of reported cases was the highest seen for decades in the USA, Australia and the UK. These are countries that provide reasonably good data. Many other countries such as Canada had a rise in 2012 also, but it did not seem to be part of a trend as in Australia and the USA. The same thing was not seen in France or Germany, but the way their data are collected makes analysis difficult. The numbers in subsequent years have not approached the 2012 level anywhere even though they have tended to remain higher than previously. Something unusual happened in 2011 and 2012 but precisely what caused simultaneous peaks is unknown. It can be added to the other enigmas of pertussis.

I have looked at the official published data for England and Wales around these years and tried to understand some of the numbers. They often change markedly in response to procedural modifications more strongly than likely changes in true incidence, and beg a big question about how much of the increased reporting is real and how much is illusory, consequent upon increased recognition and confirmation.

Public Health England regularly publishes updates on notifiable disease data and trends. For diseases such as meningitis it is a fairly straightforward task. For pertussis, the data exist, but it is accepted that their interpretation is more difficult. Not only is the disease hard to suspect and diagnose, but for nearly 20 years the methods of confirming pertussis have changed and are still changing. Looking at some of the data in detail helps to shed light on some of the phenomena I have described in this book. Not least among the difficulties of interpretation is the fact that notifications since 1940, and confirmations, since 1994, give us two sets of data, and it is not always clear which is the most relevant in any particular circumstance; indeed, pertussis is one disease where the distinction between clinical disease and laboratory confirmed infection is important, and will be even more so in the future as diagnostic methods improve and are used more. Public Health England publishes numerical

© The Editor(s) (if applicable) and The Author(s), under exclusive license to
Springer Nature Switzerland AG 2020
D. Jenkinson, *Outbreak in the Village*, Springer Biographies,
https://doi.org/10.1007/978-3-030-45485-2

data on notifications and confirmations but currently only publishes confirmations in graphical form on its website, implying that these are the most important numbers, as indeed they may be. However, the possible distortions through changed testing methods inherent in them are worth a closer look.

Incidence graphs like Figs. A3.1 and A3.2 need careful interpretation because of the way the data are presented. They are based on a population of 100,000. The 100,000 denominator relates to the age group being considered. For example, in Fig. A3.1 the blue broken line represents infants under three months of age. The left of the line starts at a level of 115. This means that in 1998, for every 100,000 babies

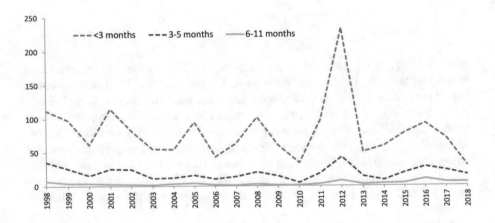

Fig. A3.1 Laboratory confirmed cases of pertussis in England 1998–2018 in under ones by age group per 100,000 population (Public Health England)

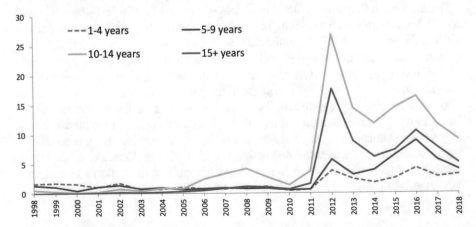

Fig. A3.2 Laboratory confirmed cases of pertussis in England 1998–2018 in over ones by age group per 100,000 population. Serology confirmation came into use in about 2006, oral fluid for 5 to 16s in 2013 and for 2 to 16s in 2018 (Public Health England)

in the population of England under three months of age, 115 had confirmed pertussis. The actual number of such babies affected happened to be 168. It can therefore be calculated that the number of babies in England at that time in that age group must have been 168/115 x 100,000, approximately 146,000.

Another example is in the green line of 10- to 14-year-olds in Fig. A3.2, in which the peak in 2012 has a value of 27. This means 27 confirmed pertussis cases for every 100,000 10- to 14-year-olds in the population in 2012. The actual number of cases was 806, so there must have been 806/27 x 100,000 in that age group in 2012, approximately three million.

Expressing the figures this way allows us to see which age group is at most risk of catching the disease and it is clearly the under three-month-olds (Fig. A3.1). In 2012 the chance of a baby that age acquiring confirmed pertussis was 1 in 435 (100,000/230). The chance for a 10- to 14-year-old in the same year was 1 in 3700 (100,000/27) (Fig. A3.2).

Under Ones

Figure A3.1 represents the under ones. By that age they should all have completed their primary course of immunisation: with whole cell vaccine until 2004, afterwards with acellular vaccine. Both types are believed to give excellent protection in the first year. The final of the three doses is given at four months but some will be delayed for a month or two for social reasons or intercurrent illness, and a few, about 5%, never receive the full course. I think it is possible that the red and green lines of the over three-month-olds represent babies that are unimmunised or partially immunised rather than vaccine failure. There seems to be little difference between incidence trends in any of the lines, except the 6- to 11-month-olds, who peak in 2016 rather than 2012 for no obvious reason, but it might be just chance because the actual numbers are very small.

The main feature of Fig. A3.1 is the 2012 under three-month-old peak that is twice the size of the peaks before and after, which are all much the same. This age group is almost always going to be hospitalised and therefore most likely be diagnosed accurately, but there is possibly a small additional number that go undiagnosed. The four yearly cycling is strikingly clear in this group which is a 'barometer' of circulating *Bordetella pertussis* in the population. It does not reflect any 'resurgence' before the doubling in 2012. The 2016 peak would probably have been equal to 2012 without the pregnancy booster that was introduced in October 2012. Uptake was 60% to 70% and it is 90% effective, so presumably influenced the statistics of under three-month-olds from 2013 onward. If the booster were given in all pregnancies the broken blue line (Fig. A3.1), already at a record low in 2018, would be expected to drop to a previously unknown low level.

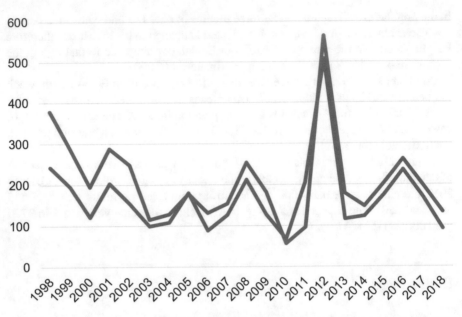

Fig. A3.3 Notifications of pertussis (blue) in England 1998–2018 and laboratory confirmed cases (brown) under one year of age

The peaks in 2012 and 2016 could be considered a small resurgence because they are the highest incidences in unimmunised under three-month-olds for at least 20 years.

Figure A3.3 compares confirmations and notifications. Notification data are published as the total under one year of age, whereas confirmations are recorded in three under one-year-age bands. Therefore, to compare notifications and confirmations, I have totalled the three groups of confirmations.

Reliable serology and PCR confirmation tests came in from 2002 and made a small impact (Fig. A3.3), visible as the separation between the lines.

In this under one age group it is possible that new tests made diagnosis easier rather than more accurate because laboratory diagnosis is more probable at this age and per-nasal swab data for culture more diligently undertaken than in older children. After 2002, notifications continue to marginally exceed confirmations. The difference is made up of purely clinically diagnosed cases.

There were 19 infant deaths from pertussis in the five years after the introduction of the pregnancy booster. All were under three months of age. The mothers of 17 of them did not have the booster. The other two had it, but too late to work properly. There could hardly be better evidence of effectiveness.

One- to Four-Year-Olds

This group is represented by the broken blue line in Fig. A3.2. These are laboratory confirmations. The line slowly drops from 1998 until 2012 when the big rise occurs, but they are then the group in this graph with the lowest incidence, telling us they are the best protected of the over ones. Their 2012 peak looks about the same as the 2016 peak. The last point on the line shows a rise in 2018 while all other age groups are falling. This possibly reflects the fact that the minimum age for oral fluid testing was reduced to two years in 2018 so it could then be diagnosed more easily without having to take a more challenging blood sample or per-nasal swab for culture or PCR.

The same age group is shown as total numbers of notifications and confirmations in Fig. A3.4. In 1998 there were ten times as many notifications because confirmation then was only by the insensitive per-nasal swab method. In 2018 there were still three times as many notifications as confirmations, probably because of the difficulty of taking blood or swabs at this age, although some were now being confirmed by oral fluid. This is not an age group in which one would expect increased recognition to be a major factor, because historically they were the group in which pertussis was most common. The rise in notifications over confirmations that is quite marked in 2008, and the even bigger jump in notifications in 2012, with a relatively small increase in confirmations, shows this age group was still being diagnosed mainly on clinical grounds. The recently increased number of cases in this age group cannot therefore be attributed to easier confirmation either. From 2008 this group all had

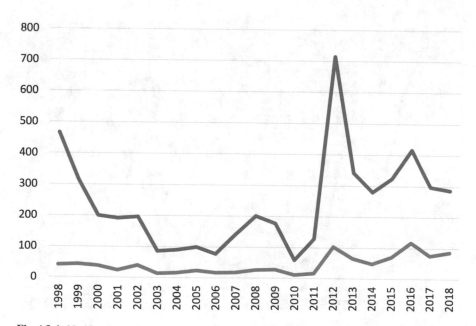

Fig. A3.4 Notifications of pertussis (blue) in England 1998–2018 and laboratory confirmed cases (brown) in 1- to 4-year-olds

primary acellular vaccine and a booster of the same. The national immunisation rate had been 93% since 1993. This looks like a resurgence in this age group but could equally be better recognition because the 2016 peak was much the same as the 1998 level just 18 years before.

Five- to Nine-Year-Olds

Five- to nine-year-olds on the red line in Fig. A3.2 show an approximately five times rise in 2012 over 2008, and ten times rise in 2016 with hardly any drop in between. This will probably be due to the introduction of oral fluid testing for 5- to 16-year-olds in 2013. The higher incidence shows they are less well protected than the one- to four-year-olds as the immunisation uptake is the same in both groups.

Figure A3.5 shows that the notification rate in the five- to nine-year-old age group is similar in 1998 and 2012. The impact of a simple oral fluid diagnostic test in 2013 shows clearly. Most of those on the right-hand side of the graph will have received an acellular primary course and booster. It is difficult to acknowledge a true resurgence when the 1998 notifications have not yet been exceeded and confirmations have risen so markedly.

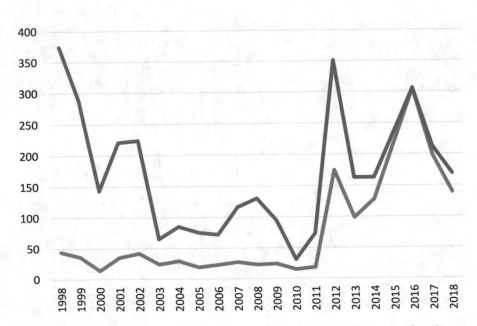

Fig. A3.5 Notifications of pertussis (blue) in England 1998–2018 and laboratory confirmed cases (brown) in 5- to 9-year-olds

10- to 14-Years-Olds

This group is represented by the green line in Fig. A3.2. With the exception of the under three-month-olds they are the most susceptible age group, with a rise of confirmations in 2006 when blood testing became more commonplace and Harnden's paper describing how common it was in teenagers was published. The apex of that rise came in 2008 and was followed by a big spike in 2012 and a slightly smaller one in 2016 with a small drop between the two. The introduction of oral fluid testing in 2013 for 5- to 16-year olds made confirmation even easier in this group.

The excess of confirmations over notifications in this group (Fig. A3.6) from 2010 suggests to me that GPs were saying to these patients, 'It could be whooping cough, but you will need a blood test. Are you willing to have it done?'

Those that agreed would often be confirmed; those that declined would probably not be notified. After 2013 a more acceptable oral fluid test could be offered but the kit was only available *after* notification, a step that I believe was not often taken.

This group shows a definite rise of both confirmations and notifications over the period. On the face of it, it looks like a resurgence but could equally be better recognition and confirmation. Most of those in the elevated right-hand section will have been immunised with whole cell vaccine.

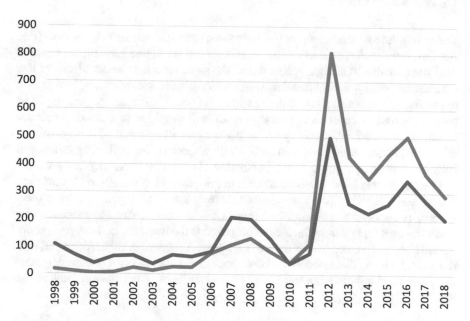

Fig. A3.6 Notifications of pertussis (blue) in England 1998–2018 and laboratory confirmed cases (brown) in 10- to 14-year-olds

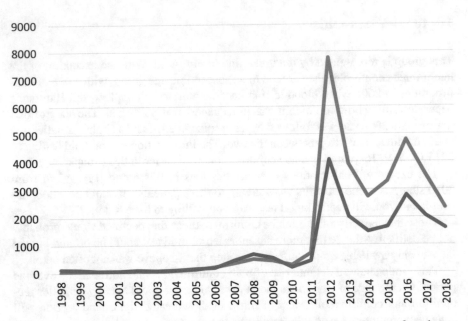

Fig. A3.7 Notifications of pertussis (blue) in England 1998–2018 and laboratory confirmed cases (brown) in 15-year-olds and older

15 Years and Older

The purple line of this group could only ever have had whole cell vaccine (Fig. A3.2). It is the same shape as the 10- to 14-year-old green line but a bit lower and it does not have the 2008 high. It also shares the same reversed characteristics of low notification and high confirmation (Fig. A3.7) for the same reason, which is a bypassing of the notification system in many cases. A GP managing a suspected case of pertussis, almost always now an adult, will simply send a blood sample for testing. If it is positive, probably no further action will be taken. If it is negative it will probably not be notified, even if clinically suspected. The bulk of cases numerically are in the purple group as the population of over 15s is considerably greater.

Once again this looks like a resurgence in this group as in the 10- to 14-year-olds. The reason I have doubt about whether there is a true resurgence in these older groups, as opposed to unimmunised infants where there is stronger evidence of a small resurgence, is that in the Keyworth practice the numbers in these age groups started high in 1986 and have remained high (Fig. A2.2) with a slight recent fall until the present day. As discussed in the book, doctors did not believe it occurred in those age groups and it therefore never got diagnosed until easy confirmation came along.

*****.

Reference

1. https://assets.publishing.service.gov.uk/government/uploads/system/uploads/attachment_data/file/797304/Laboratory_confirmed_cases_of_pertussis_in_England_2018.pdf

Index